Citizenship, Activism and the City

Were the occupations of 2010–11 – from Spain to Tahrir Square to Occupy Wall Street – a success or failure? Are they the model for urban radical politics? This book challenges common understandings and underlying assumptions of what constitutes activism and resistance. It proposes a critical urban theory of politics and citizenship that is grounded in the city as it is inhabited. For those who are marginalized, the city is a double-edged sword of oppression and emancipation.

This book argues for an intersectional approach that actively dismantles hierarchies and embraces a wider range of acts of resistance and creative transformation, one in which we recognize these acts of citizenship as a form of constitutionalism. Wood reframes the theorization of protest and of the city, 'post-political' literature and the history of protest, and Marxist and anarchist ideas about the time and space of politics. Through this, she adopts a unique approach to provide new theoretical insights and challenges to post-political thinking.

This book will be valuable reading for those interested in political, urban and social geography, in addition to political economy and progressive politics in the urban context.

Patricia Burke Wood is Professor of Geography at York University, Toronto, Canada.

Routledge Research in Place, Space and Politics

Series Edited by Professor Clive Barnett, Professor of Geography and Social Theory, University of Exeter, UK.

This series offers a forum for original and innovative research that explores the changing geographies of political life. The series engages with a series of key debates about innovative political forms and addresses key concepts of political analysis such as scale, territory and public space. It brings into focus emerging interdisciplinary conversations about the spaces through which power is exercised, legitimized and contested. Titles within the series range from empirical investigations to theoretical engagements and authors comprise of scholars working in overlapping fields including political geography, political theory, development studies, political sociology, international relations and urban politics.

For a full list of titles in this series, please visit www.routledge.com/series/PSP

Architecture and Space Reimagined: Learning from the difference, multiplicity, and otherness of development practice
Richard Bower

Migration, Squatting and Radical Autonomy: Resistance and Destabilization of Racist Regulatory Policies and B/Ordering mechanisms
Edited by Pierpaolo Mudu and Sutapa Chattopadhyay

Nation Branding and Popular Geopolitics in the Post-Soviet Realm
Robert A. Saunders

Political Street Art: Communication, Culture and Resistance in Latin America
Holly Eva Ryan

Geographies of Worth: Rethinking Spaces of Critical Theory
Clive Barnett

Psychological Governance and Public Policy: Governing the mind, brain and behaviour
Edited by Jessica Pykett, Rhys Jones and Mark Whitehead

Citizenship, Activism and the City: The Invisible and the Impossible
Patricia Burke Wood

Citizenship, Activism and the City

The Invisible and the Impossible

Patricia Burke Wood

Routledge
Taylor & Francis Group
LONDON AND NEW YORK

First published 2017
by Routledge
2 Park Square, Milton Park, Abingdon, Oxon OX14 4RN

and by Routledge
711 Third Avenue, New York, NY 10017

Routledge is an imprint of the Taylor & Francis Group, an informa business

British Library Cataloguing in Publication Data
A catalogue record for this book is available from the British Library

Library of Congress Cataloging in Publication Data
A catalog record for this book has been requested

ISBN: 978-1-138-74680-0 (hbk)
ISBN: 978-1-315-18042-7 (ebk)

Typeset in Times New Roman
by Taylor & Francis Books

MIX
Paper from responsible sources
FSC www.fsc.org FSC® C013604

Printed and bound by CPI Group (UK) Ltd, Croydon, CR0 4YY

For Madigan and Sean

Contents

Acknowledgements

I would like to acknowledge the inspiration of two books, which I read more than two decades apart: Claudia Rankine's *Citizen: An American Lyric* (2015) and Patricia Williams' *The Alchemy of Race and Rights* (1991). These books, both written by Black American women scholars (one in English Literature and one in Law), are two outstanding examples of genre-busting scholarship that integrates the personal and the political, the grounded and the theoretical. Their voices speak with and from an engagement, embodiment and humanity that is commonly absent from academic scholarship. The results are both grounding and sky-opening in their insights. I do not pretend for a moment that my own work achieves the same heights, but I am deeply grateful for the space they carved out in academia.

I wrote half of these pages in Robarts Library at the University of Toronto during a sabbatical leave from my teaching at York University. I am grateful for the time, and for the public access to such a beautiful and quiet space to work. (Yes, this is a shout-out for the importance of public libraries.) I wrote the other half in Dublin, Ireland. Heartfelt thanks go to Nick, Deirdre, Christopher, Andrew and Jacob McGivney for their friendship and extraordinary hospitality. Living with all of them was a gift. I also want to thank the staff at the Club na Múinteoirí for providing me with the perfect place to work, as well as all the tea I could drink.

There are many more people without whom I would not have finished nor even started this book. Its limitations and failings are entirely my responsibility, but the work was eased and improved because of their contributions.

I must offer special thanks to my writing buddy, Matt Farish, for helping me stay on track, for being a 'first responder' to early drafts, and for supporting the idea of the book before I started writing it. Ranu Basu understood early on what I was trying to do, and helped me do it better. Kim Michasiw, Ranu Basu, Richard White, David Rossiter and Brenda Fine all read drafts and gave me valuable feedback.

I had two early opportunities to develop the ideas for this book, and for this I would like to thank Julie Young and her fantastic graduate class at McMaster University, and Alison Bain, past-Director of the City Institute at York University, and those who attended my talk in January 2016.

To the extent the idea of writing this book has a precise starting place, it was a conversation over tea with Aubrey Robinson at Backstage in Dublin, in December 2011, and I am grateful for that spark and our friendship. Other friends and colleagues in Ireland who have provided support, encouragement and inspiration include Patrick Burke and Emma Bolger, Dave Johnson, John Bissett, Joan Murphy, Karl Deeter, Simon Farrell, Damien Peelo, Bernard Joyce, Colette Spears, RoseMarie Maughan, Donnacha O Briain, Mary Gilmartin, Eoin O'Mahony, Kylie Jarrett, Mark Boyle, Cian O'Callaghan, Elizabeth Mathews and Fiona DeLondras.

Mary Gilmartin and Kylie Jarrett provided pep talks at critical moments, for which I give special thanks. Thanks to Jess McLean, for the music suggestions.

While I was working on this book, Richard White and I co-edited a special double issue of the *International Journal of Sociology and Social Policy* (*IJSSP*) on activism with and without organization, based on sessions we organized for the Chicago American Association of Geographers' meeting in 2015. Hearing the conference presentations, and then reading and editing the *IJSSP* articles stimulated my thinking significantly. Most of all, ongoing conversations with Richard have led me to more literature and ideas, and I am grateful for that friendship.

At home in Toronto, several others provided friendship, support and inspiration: Colleen Burke, David Rider, Shawn Micallef, Scott Prudham, Randy Katz, Deirdre Smith, Liette Gilbert, Susan McGrath, Barbara Rahder, Linda Peake, Alison Bain and Cathy Clarke, and Deborah McGregor. Special thanks to my fellow recovering administrators, Tom Loebel, for the extended conversations about psycho-analytic theory and politics, and Kim Michasiw, for the gift of ergative verbs.

Many other friends and colleagues have shaped the thinking that went into this project, though some remain blissfully unaware of this. They include Engin Isin, Evelyn Ruppert, David Rossiter, Mike McQuarrie, David Hugill, Nik Heynen, Mark Purcell, Brian Ray, Simon Springer, Nathan Clough, Anthony Ince, Vanessa Sloan Morgan and Joseph Nevins.

Faye Leerink at Routledge was encouraging from the very beginning, and the final steps of getting the book into print were eased by the skilful assistance of her colleagues, Priscilla Corbett, Richard Skipper, Kris Wischenkamper and Matthew Twigg. Thanks also to Elinor Fitzsimons for her advice with special permissions and to Margaret de Boer for compiling the index.

I want to thank my parents, who have always taken an interest in my work and made it better. As a single parent, I literally could not do my job without their logistical and emotional support. Just as I started writing, my dad listened to me talk out this book over dinner on a trip to New York, and my mom sat patiently through rambling iterations over tea, countless times. My extended family also put up with me as I did some writing during a family vacation on Georgian Bay, so thanks for that, too.

While I was writing, the tragedies that give rise to the politics I am trying to highlight and understand occurred almost daily: the fire at Carrickmines,

Dublin, which killed ten members of two Traveller families, including five children; the massacre in a gay and predominantly Latinx nightclub in Orlando; the shootings of Alton Sterling and Phaidon Castile by police in Louisiana and Minnesota; the death of Abdirahman Abdi, who was beaten by the Ottawa police; the shooting of Colten Boushie of the Red Pheasant First Nation by a man he hoped would help him with his car – these are but a few. Writing this book made me cry, from grief and sadness, and from awareness of my privilege and good fortune.

I am most grateful for the support and patience of my children, Madigan and Sean, and I offer them my deepest love and appreciation. They have taught me about what it means to be human, and how hard and beautiful that is, more than anyone else ever could.

Introduction

The invisible and the impossible

Concrete life

> Occasionally it is interesting to think about the outburst if you would just cry out –
> To know what you'll sound like is worth noting –
>
> (Rankine 2015: 69)[1]

Along Dublin's Dame Street, most buildings are set flush with the pavement and the road, but the Central Bank of Ireland is set back from the road, with a small plaza wrapping around the corner of Dame Street and Fownes Street Upper, at the edge of the Temple Bar neighbourhood. Today, Temple Bar is known for its bars and restaurants, packed with university students and tourists. When construction of the Central Bank building began in 1972, just as Ireland joined the European Common Community, Temple Bar was considered a slum, and the City planned to redevelop it with a bus terminal. By the time the bank building was completed in 1980, Temple Bar was revitalized with small businesses, and protests stopped the bus plan. In the early 1970s, Dublin was just beginning to transform from the capital town of a mostly rural, 'underdeveloped' economy into a global city that during the Celtic Tiger boom years of 1994 to 2008 would come to host head offices for Google and Citibank.

The Central Bank's architecture is unlike that of its neighbours. A bit of a marvel, the building's eight storeys are suspended by steel trusses from the roof, rather than resting on top of a foundation. Its concrete blocks tower over its neighbours, and share more with brutalism than with the eighteenth-century styles of the surrounding cobblestoned streetscapes or the Georgian fronts of nearby banks on College Green. Its height was a violation of its planning permission and an irritant to some members of the public and Dublin City Council; a challenge stopped construction on the building for almost two years between 1974 and 1976 (Sisk 1987). The main entrance sits at the top of a short flight of stairs with flat narrow ramps under and between the bannisters, as part of the public plaza actually extends under the main floor; in the building's early days, this area was a popular site for skateboarders. The City and the Bank took a dim view of such activity and a fence was erected between the building and its plaza, complete with 'Skateboarding Prohibited' signs.

It is hard to find public space to occupy in Dublin. Although the Republic of Ireland's independence from the United Kingdom was predicated, in part, on the street battle of the 1916 Easter Rising, the layout of Dublin does not easily accommodate public gatherings for protest. There are rallies in front of the Central Post Office, more for its symbolism from the Rising than anything, but this means standing in O'Connell Street and shutting it down to traffic. There is no central public square, no Zócalo or piazza. The country's legislature is in central Dublin, but the small courtyard in front of its buildings on Dawson Street is fenced off from the pavement, with security standing guard. The President's residence and offices are well away from the city centre. There was once a thriving market square in Smithfield Fair (the scene of the opening shot in the film, *The Commitments*, if you want to go look), on the northside, within walking distance of the city centre, but the area has since been gentrified and the market is allowed only once a month. The police regularly ask people who are sitting in the renovated play area to move along.

The choice of the plaza in front of the Central Bank was thus a good one for Occupy Dame Street (ODS) set up its encampment on 8 October 2011, for the physical space it provided, for its centrality and public visibility, and for the symbolism of the site. But as can been seen from even a brief history, ODS was only the latest of a host of contests over the site, as one finds for almost every bit of city space, big or small. Although the threat of being closed down by the police hovered over ODS throughout its existence, there was a steady public tolerance for the protest. The encampment remained in place until March of the following year, when the City announced that the route of the St. Patrick's Day parade along Dame Street necessitated clearing the plaza. As excuses go, it was pretty thin, but it didn't really matter; the protest had fractured internally by then by disputes among campers, and between those who felt the camp was necessary and those who did not. By the end of the winter, very few campers remained in the plaza. The police raided and dismantled the camp on 8 March 2012.

Occupations, from Spain to the Arab Spring to Occupy, captured the attention of activists, academics and the media in 2011 as something 'new' and as the apparent 'start' of major political and social change in the wake of the global financial crisis of 2008–09. What is this politics of mass sit-ins? How is it connected to other politics, the geography of the city, and urban life? There were other post-crisis actions and mass mobilizations, before and after Occupy, some of which became international or global movements, but which did not receive the same kind of attention. Why did some scholars and activists greet these occupations as the start of an important political resistance, but did not grant the same status to or get involved with other actions or mobilizations? Which, or whose, priorities did Occupy and similar protests reflect and produce?

Some specifics of Dublin's story are unique to Dublin, but it shares a pattern of urban politics and citizenship with other Western and non-Western cities that were built and rebuilt by global flows of capital, and whose citizens thus experienced to varying degrees the impact of the 2008–09 crisis. Some of their residents benefitted from deregulated markets and the financialization of everything; some residents protested the inequality of the good years, and of the austerity programmes that followed the crash. Many residents fall into both categories. Ireland was hit especially hard by the crisis and subsequent bailout by the Troika (European Commission, European Central Bank, International Monetary Fund (IMF)), and ODS was one of a series of protests that had taken place in its wake. Students marched in 2008 (15,000) and in 2010 (40,000) to protest fee increases to tertiary education (RTÉ News 2010). The main public sector union, Impact, organized a day of action in February 2009, and the Irish Congress of Trade Unions organized another which brought out 150,000 protesters across eight cities in November 2009, to launch an advocacy campaign against the forthcoming national budget. For a few years, starting in December 2010, community groups organized an annual 'Spectacle of Defiance and Hope', a march and rally that included the presentation of a book of grievances. In different community groups around the city, many individuals, including children, artistically set out their concerns for inclusion in the book, which was presented to the City at the culminating rally in front of the Central Post Office on O'Connell Street.

Yet there was also a sense among many residents, activists or otherwise, that the Irish were too compliant, that there was insufficient outrage directed at the bankers and developers who had gambled with the country's economy, and at the government, which insisted that the burden of recovery be 'shared' by cutting social and community programmes. A small group of women in Dublin, who were originally from Spain and were inspired by the 15-M movement in their home country, dreamed up the possibility of an encampment in front of the Central Bank, and got some other activists from housing organizations to launch an open call, which went out on Facebook and Twitter. In response to the call that it was 'time to do something', Occupy Dame Street mobilized a small and diverse group to erect tents along the line of fence in front of the Central Bank's stairs. It grew quickly to a regular group of a few dozen who spent time at the camp most days and more on weekends, to discuss what was going on and deliberate over what interventions might be possible. The group held regular meetings, planned other protests (from pickets to performance art interventions), ran a camp kitchen, organized an 'Occupy University' with public lectures by academics and other professionals (full disclosure: I gave one of these). There was often music; even Billy Bragg came and sang when he was in town.

From one angle, it was all a performance. As a performance, it was an effective strategy to gain space in the media's attention, whose pages and airwaves had been overtly allied with the government's 'necessary' austerity plans (Mercille 2015). Occupy Dame Street as a spectacle, as a physical interruption of the city's space and patterns, became an event that the media had to cover and fit into the narrative of the boom, crash and 'recovery'.

Even there, the protesters were disruptive, refusing to identify anyone as a 'head' of the movement or to articulate specific, 'doable' political goals. It was not immediately clear to the general public who these people were, what they were doing or what they wanted.

The protesters were not the global elite of Dublin. They were university students and faculty, community organizers, social workers. The camp also drew in people experiencing very hard times: those who were unemployed, homeless, dealing with addictions, or mentally ill. They were not politicians, nor union leaders. Indeed, a common refrain of camp meetings was the debate about relationships with politicians, political parties and unions. The general feeling was that they were not welcome, and that any participation was contingent on parking partisanship and agendas at the door. And there were organizers but no leaders, no one elected to speak on their behalf, no one who formally represented the camp or the movement. In these aspects, they were a similar cohort, with similar politics, as their fellow Occupy protesters in New York, Toronto, or Rome. But what did Dublin have in common with these other 'Occupy' events, really? We have become inclined to link similar but disparate events together into global movement, at the expense of the local context that made each possible. This is not a book about Dublin or Ireland, but it does aspire to grapple with contemporary urban politics, citizenship and activism in a way that would make sense to Irish people, and others on the 'margins'. ODS was one of many acts of resistance, public and private, large and small, that reflected Ireland's experience of the crisis as well as other ongoing activism that preceded it.

The global financial crisis had specific effects that are relevant here. As elsewhere, Ireland's crisis was in part the result of inflated housing markets, which had been financialized and, as a share of total gross domestic product (GDP), became a significant means of 'wealth creation' (Aalbers 2008; Lees et al. 2008; Gotham 2009; Christophers 2011). This property development and speculation had created a particular city, whose revenue streams under neoliberal policies for infrastructure had been reduced by central governments. Responsibilities were nevertheless downloaded, and the need for funds turned cities to the global market for cheap credit. Municipal bonds were commonly held in other countries. Even in countries that were spared the worst of the crash, due to having banks with less exposure, such as Canada, these processes were still intensifying, gentrifying and indebting the urban landscape (Walks 2013, 2014a, 2014b).

The crash was politically significant because it was an obvious wobble of the global financial system, and thus a potential entry point for change. There were expectations in many quarters that governments would fail, that politics would fundamentally change, that financial institutions and their transactions would be subject to new regulation – and that these reforms would come about largely from popular mobilization, including a resurgence of organized labour. Some protested the crash itself – their target was thus Wall Street and its equivalents – and some protested the austerity that followed, and their target was government and organizations like the IMF and the EU. The crash exposed the fragility of the system. For some, it was due to internal contradictions; for others, it was evidence of corrupt and greedy individuals, bad apples who had spoiled it for everyone. Both were true in Ireland.

The actual activist politics that emerged in the wake of the global financial crisis, however, have challenged scholarly and community expectations. Peter Marcuse observed in a 2011 publication (likely written just before Occupy), 'Yet the protest has been subdued. Life on the campuses barely notices the crisis. Labour unions confine themselves to asking for limits on CEO pay' (Marcuse 2011: 29). Urban social movements seemed to mobilize very slowly and not very publically; new movements formed in unpredictable ways and from unpredicted sources; even the global phenomenon of 'Occupy' sprung up around the world mostly without coordination by or even affiliation with existing activist organizations. The relationship of post-crisis urban activism, particularly Occupy, to organized labour is not straightforward. Some labour unions explicitly did not support the Occupy movement. According to Nathan Schneider's account organizing Occupy Wall Street, 'when people from the General Assembly tried to reach out to unions and the like, none wanted to touch the occupation with a ten-foot pole' (Schneider 2013: 18; Graeber's account is more positive – see Graeber 2013). Many labour activists complained (not inaccurately) that Occupy drained energy, attention and bodies from other, on-going campaigns and activities, although in many instances the relationship was more complicated (Tufts and Thomas 2014). As austerity gutted community resources in many countries, communities responded in highly individualized ways, on a variety of scales, that did not fit many schedules or expectations.

The crisis hit different social groups differently, and was geographically uneven at every scale. Those who had stretched themselves the farthest to purchase a home were more likely to find themselves facing foreclosure, but job losses wreaked havoc even with those who thought they were financially secure. Already vulnerable households found themselves more so. Community programmes that were cut in Ireland affected low-income groups and Irish Travellers more than others. The United Kingdom's choice of post-crisis austerity programme gutted public housing, and left disabled people with fewer critical supports. In Spain and Greece, the youth unemployment rate skyrocketed. In the United States, Cohen identified African-American youth as 'the group hit hardest by the economic recession that started in 2008' (Cohen 2010: 8). David Graeber argues that this is, in part, the effect of neoliberalism as a primarily political project that seeks to demobilize communities, rather than the economic project it claims to be. David Harvey has similarly argued neoliberalism's primary goal was to incapacitate labour (Harvey 2016a), and Jamie Gough and Byron Miller have suggested it seeks to destroy the social (Gough 2002; Miller 2006; see also N. Rose 1996). Post-crisis policies often intensified existing neoliberal practices (Peck et al. 2012), rather than reversing them, and the discourse of economic logic remained relatively undisturbed: 'the first reaction to the crash was not – as most activists, including myself, predicted – a rush towards Green Capitalism, that is, an economic response, but rather, a political one' (Graeber 2011: 113). Given the economic decline produced by austerity measures that cut spending in a recession, the more likely justification is a political agenda (Hearne 2015; Graeber 2011).

Some of our misplaced expectations and failure to grasp what was happening in urban activism is due to our focus on the large public occupations and, especially in North America, a fixation on Occupy. Just as Occupy distracted attention from other political movements, much of the work on Occupy in particular has focused on NYC or the US, with less investigation into the diversity of sites; Occupy alone was a 'movement' with an estimated 1500 sites. Following Tahrir Square, Spain and Occupy, occupations were re-established as the model of protest, even as it would be difficult to identify what the attributes of that model were. Much of its importance derives from those who announced it as important. In its scholarly isolation, its relationship to movements such as the Arab Spring, Black Lives Matter, Idle No More and anti-austerity campaigns in Europe remains undertheorized, as well as other pre- and post-crisis protest, which is just beginning to emerge in the scholarly literature. For example, in many ways, Black Lives Matter is a specifically post-crisis movement. The school-to-prison pipeline, the racialized policing of the city, the white flight from cities and subsequent gentrification of the same cities to bring whites back – impoverishing Black families at each turn, carding and stop-and-frisk practices, the gutting and privatization of the urban public school system: these are also the direct results of neoliberalism (Hackworth 2006), as the financialized city became the object and means of wealth extraction. As I write this, less than a decade has passed since the crisis, and both government policy and resistance continue to unfold, newly complicated in the EU by the Brexit decision.

Some public and scholarly considerations of early twenty-first-century activism and protest have focused on the digital technologies used to organize, mobilize and educate. Much has been made of their use of digital media, especially social media. Occupy is one mobilization among many, and one of the functions of the critical emphasis on social media has been to de-differentiate, although dependence on social media is arguably a different thing in Cairo than it is on Wall Street. Manuel Castells argues (somewhat naively, but we will come back to that) that the Internet is key to the organization and mobilization:

> It began on the Internet social networks, as these are *spaces of autonomy, largely beyond the control of governments and corporations* that had monopolized the channels of communication as the foundation of their power, throughout history. By sharing sorrow and hope in *the free public space of the Internet*, by connecting to each other, and by envisioning projects from multiple sources of being, individuals formed networks, regardless of their personal views or organizational attachments. … From

> the *safety of cyberspace*, people from all ages and conditions moved towards occupying urban space, on a blind date with each other and with the destiny they wanted to forge, as they claimed their right to make history – their history – in a display of the self-awareness that has always characterized major social movements. The movements *spread by contagion in a world networked by the wireless Internet* and marked by fast, viral diffusion of images and ideas.
>
> (Castells 2012: 2, emphasis added)

In his study of several 2011 movements (Tahrir Square, Madrid and Barcelona, Occupy Wall Street), Paolo Gerbaudo has noted that the 'uses of social media among activists are as diverse as their venues' (Gerbaudo 2012: 3), and argues they are a tool of mobilizing participants into the streets, to '*choreograph* collective action', similarly to previous communication tools, such as newspapers and leaflets, but even less hierarchically (4). Both Castells and Gerbaudo note the open, horizontalist manner of organizing, but Gerbaudo especially seeks to insert emotion and embodiment into our analysis and understanding of protest and social media, and underscores the importance of 'emplacement' and the ways in which these movements are quintessentially urban, and of the city. Engin Isin and Evelyn Ruppert (2015) have similarly emphasized the urban character of the 'digital citizen', whose online interactions possess the same anonymity, speed, and vulnerability to surveillance as life offline.

For many, these events are part of a larger, longer conversation about the possibility for a 'new Left', following (variably) the collapse of Soviet Union and socialism, and the rise of neoliberalism and austerity programmes. In this 'post-political' context, where 'the end of history' alleged to place major political and economic questions beyond debate, building solidarities to mobilize for alternative futures has been frustrated and outmaneuvered by managerialism and 'common sense' politics. However, we also need to challenge the limits of the work that has critiqued post-political thinking. The otherwise excellent work regarding the foreclosure of political knowledge and voice has yet to fully engage with work that moves beyond 'the view from nowhere', and has yet to fully incorporate intersectionality (Crenshaw 1989) and the political experiences to which the latter literature gives voice. What the post-political critique usefully highlights are the ways in which neoliberal urban development has foreclosed discursive and material space for politics. The idea of an 'impossibility' of politics where 'there is no alternative' as a *new* phenomenon, however, ignores the history of the politics of women, indigenous peoples, Blacks and other racialized groups, LGTBQ communities, and many others – and the history of the city they have inhabited. Their – *our* – politics have always been labelled impossible. There was (and in many cases still is) no alternative to settler colonialism. No alternative to slavery. No alternative to patriarchy.

These perspectives of the neoliberalized city and its marginalized communities need to be brought together. For example, newly arrived immigrants used to live in the centre of town, frequently in slum conditions, but at the heart of activities they could disrupt. Today, wealthy immigrant communities buy homes in the suburbs, and immigrants who arrive with little are in inner-city suburbs or the 'in-between city', where accessible public spaces for social gathering or political activity are few and far between (Sieverts 2003, Young et al. 2010). The State may well have abandoned these areas, but this has not erased their politics or community-building (Wood 2013). Their strength is in their local neighbourhoods, in 'third places' of private businesses that serve as community hubs for a multiplicity of purposes – social, practical and political (Oldenburg 2001). As Henri Lustiger-Thaler puts it, in the fragmented city, community is the localization of power (Lustiger-Thaler 1994). Their public life is at the barbershop, in the salon, at the doughnut shop, at the playground, in the local library, in the pub: they share information, political opinion, employment opportunities, advice for youth (a shadow network, perhaps, of the elite's golf club spaces) (Wood and Brunson 2011). They 'get things done' by helping each other, by looking after each other's children, sharing rides, sharing food (see Cohen 2004). That is a politics – more than that, it is a political economy: the social and political relationships and processes that determine the distribution of goods and services among the community (Gibson-Graham 1996). Žižek in particular 'depoliticizes those acts of becoming political by restricting the political to those "revolutionary" actions that seek universal restructuring' (Isin 2002: 277). We must not overlook these acts, despite their banality or small scale. How, then, can we understand such acts as the politics that they are?

Fundamental to the specific consideration of Occupy and post-crisis activism, and to a broader reconsideration of politics and the post-political, is a re-engagement with that inhabited city where politics is also exercised in informal spaces and practices. Post-crisis protest is directly related to the city in several important ways. From a 'top-down' perspective, the reshaping of the city under neoliberal development, and the uneven and inequitable impact of austerity budgets on the city's economy and landscape have made more visible the hardship, loss, surveillance and lack of political voice for its less privileged residents. These visible symptoms created catalysts for, and even sites of, protest, even as that protest was made more complicated by the policing and privatization of public space. Moreover, from the perspective of the 'bottom-up' city, the resistance has its roots in an alternate geography of the city as people inhabit it, which exposes the incompleteness of capital's grip on urban life and form. This city has been and remains emancipatory, where residents seize the opportunities that the diversity, anonymity, fluidity, and plasticity of the city also provide. In order to understand post-crisis activism and protest, it is particularly this latter, social city of actual bodies that we must bring to the surface again and integrate with our political-economic theorizations of the city.

We need to understand the city of these politics, too: a city formed by capital but not just by capital – by patriarchy and sexism and homophobia, *and* by resistance to all of them. Paradoxically, the city at times provides an accessible, visible stage for that resistance (Holston 1998). It is important to recognize that this also produces the city; it disrupts other flows (Lefebvre 1991 [1974]) but also stands outside of them, producing mobilities, exchanges and spaces out of relationships and desires, independent of markets. We must start with the lived experience of that struggle. As Lefebvre insisted, 'The specific, embodied, lived experience of the city and the struggle for it remains at the heart of such analysis, which must be able to "grasp the concrete"' (40). The concrete is not merely the tangible; in Lefebvre's view, our realities consist of all of the 'physical, mental and social', always in 'trilectic' relationship with each other, and they are also a product of what has actually happened. 'If space is produced…then we are dealing with *history*…' (21, 46).

As Castells emphasized, 'The definition of urban meaning will be a process of conflict, domination, and resistance to domination, directly linked to the dynamics of social struggle and not to the reproductive spatial expression of a unified culture' (Castells 1983: 302). The city is fundamentally a site of struggle (Deutsche 1996; Isin 2002; P. Wood 2014) and a place of oppressive surveillance and control. 'There is nothing more contradictory than "urbanness"', Lefebvre wrote. The city hosts 'the setting of struggle', 'the stakes of that struggle', and 'the places where power resides' (Lefebvre 1991: 386). The streets are and have always been dangerous for many individuals. The struggle continues and shapeshifts under neoliberal capitalism. In addition to increased State and police surveillance, we also have corporate forms of surveillance through social media, QR codes, and mobile phone tracking. New forms of governance we cannot always see, along with the many ways in which we discipline ourselves to survive and succeed in the neoliberal city, have changed the shape of citizenship. As Proudhon argued in 1848, 'To be ruled is to be kept an eye on, inspected, … listed and checked off…' (cited in Scott 1998: 183). Governance is a daily, routine aspect of the city, woven through public and private space, but it is unevenly applied. The movements of women, Blacks, lesbians and gay men, and trans people are monitored and policed in different and frequently more violent ways than those granted more privilege.

At the same time, many marginalized individuals and communities have also sought and found safety and liberty in both public and hidden spaces of the city, in ways that are unavailable in rural communities or in the private space of home (P. Wood 2014). They have also found moments and places in which to openly challenge their oppression. The emancipation of the city is deep in its history – as a place where you could literally purchase your freedom (Isin 2002: 10–11). This understanding of the city is a necessary complement to our political economy frames. Capital's successful exploitation of the city relies on, but also struggles to contain, the disruptive difference of the city: the diversity, fluidity, anonymity, flexibility, creativity, innovation, and cheap labour of the concrete life of ordinary and marginalized people.

These perspectives on the double-edged sword of city life and its agonistic politics are still largely missing from urban and political writing that is otherwise critical and potentially sympathetic. Critical urban theory, particularly with respect to the 'right to the city', neglects the diversely inhabited city in its theorizations and the difference that diversity makes to theory and practices of politics. Lefebvre argued eloquently that the city was now the political equivalent of the factory floor, that capitalists were using and shaping the city for purposes of profit-seeking and capital accumulation – and not just through its built form, but even social relationships within it. Moreover, the very idea of space – how it could be ordered and planned – was fundamentally produced by capital, or capitalist thinking. This is well rehearsed among urbanists and has been built upon by David Harvey (1985, 2013), Neil Smith (1996), and a host of other (mostly Marxist) scholars. As Brenner, Marcuse and Mayer have written, 'Cities, [Lefebvre, Castells and Harvey] argued, are major basing points for the production, circulation, and consumption of commodities, and their evolving internal sociospatial organization, governance systems, and patterns of sociopolitical conflict must be understood in relation to this role' (Brenner, Marcuse and Mayer 2011: 3). In parallel to the capitalization of the urban is the urbanization of capital, as the processes of capital accumulation, surplus, consumption and reinvestment were wired through both cities and newly networked and monitored places, and all space has thus become increasingly 'urban' (Lefebvre 1991; Harvey 1985; Brenner 2012). This is brilliant scholarship that illuminates cities as not only places or sites, but as processes and as spaces of flows. It is fundamental to unpacking the neo-liberal city, with its intense financialization and privatization of housing and other real estate that played a starring role in crashing the global economy in 2008–09. It remains essential to understand how the city is formed, built, and exploited by capitalists and the politicians who accommodate them.

Nevertheless, this understanding of the city is partial, and as a basis for critical urban theory and any radical politics of resistance, it will not suffice. In a 2011 essay (a condensed version of a larger talk), Harvey notes that 'class struggle in and around the urban is as important as class struggle in around the factory' and 'they need to be unified' (Harvey 2011: 267). However, this (longstanding) incorporation of city space has not changed the composition of a politics of resistance. Harvey asks 'who produces the city', and his own answer is 'workers', not residents. Moreover, although he acknowledges in passing the supportive role of others, his concept of work includes only paid labour, and class consciousness is his only basis for political struggle. The political subject is not the citizen, by formal or informal definition; it is the employed, class-conscious worker. That particular consciousness of that particular struggle with the exploitation of capitalism is what constitutes politics.

The 'right to the city' and critical urban theory more broadly must include, for starters, not just workers but all residents, and it must address more than the political economy of global capitalism. As Mark Purcell noted in one of his contributions to the much-debated, perhaps-often-misunderstood idea, Lefebvre's argument for the 'right to the city' was not to privilege the urban as a scale, but to argue that the rights of inhabitants should not be diminished or extinguished in favour of owners (Purcell 2006). Before we can argue for reclaiming our right to the city, we need to incorporate a fuller understanding of that city and all its inhabitants.

Purcell's work incorporates Lefebvre's categories of 'industrial city' and '*l'inhabiter*', or the inhabited 'urban society'. In the industrial city we have 'a city in which private property and exchange value are the dominant ways to organize urban space'. This produces a spatially and socially fractured city. The different parts of the city are allocated values for market exchange that reflect their physical locations and structures, as well as a social assessment of residents. These two valuations influence and increasingly determine one another. Residents are encouraged to be consumers rather than intervening citizens in these processes. Purcell concludes, 'It is, in short, an oligarchy, a city managed by an elite few state and corporate administrators. Today we would call this the neoliberal city' (Purcell 2013: 22). On the other hand, in Lefebvre's urban society, the management of urban space is in the hands of residents who have the ability to claim and make use of space as they see fit. Where the industrial city is 'actual', urban society is 'virtual'. Virtual here does not mean 'a utopia, a no-place, an ideal imagined out of the ether that can never exist', but rather 'an extrapolation or amplification in thought of practices and ideas that are *already taking place in the city*', but have not yet realized their potential (Purcell 2013: 23, emphasis original). To achieve radical democracy under these circumstances, Purcell starts with Lefebvre's idea of 'transduction': 'a way to cut a path that leads beyond the actual world already realized and toward a possible world yet to come' (21). Purcell (along with Lefebvre) is more hopeful than I am, and I start in a darker place, but I think we are walking in the same direction. I find the difference between the industrial city and *l'inhabiter* to be an indistinct one (as I shall address farther along), but nevertheless, for Lefebvre and Purcell, and for me, this project of radical politics is specifically located in the inhabited city.

Post-crisis activism is something 'new', in that it is in many ways the product of the politics of neoliberalism and the welfare state, of post-industrial urbanism and global financial centres, and the collapse of 2008–09. However, we need to think through what mass occupations are and are not, to put local movements in their global contexts, and also rip them free of global generalizations. It also means looking beneath the mass occupation movements, and tracing their connections to other politics, and consider the possibility, for example, that Occupy might have more in common with a flash mob than a general strike. We must ground this work in the concrete life of the inhabited city. If we think of the city's residents as political subjects resisting forces that govern them in some way, then we are speaking of citizens. Underlying their actions, their discourses and our analyses of them is a fundamental question of citizenship, of what it means to be political.

As Isin has argued, the 'intense struggles, conflicts, and violence to wrest the right to becoming political from dominant groups, which have never surrendered it without struggle' are what determine the shape and breadth of citizenship (Isin 2002: 2). This manifests itself in the city, because while 'the dominant groups have never been inclined to give an account of their dominance' (5), Isin challenges the elite presentation 'that the city was always a *unified* agglomeration of tribes "settling together"' (6). In fact, the city is characterized not by its unity but by its diversity, and thus, it 'is the battleground *through which* groups define their identities, stake their claims, wage their battles, and articulate citizenship rights and obligations' (50). Its elites have continually sought to create and present a unified, ordered, efficient city. In this light, for example, James Scott has critiqued Le Corbusier's Paris as the aspiration of architecture to overcome inefficiency, to make the city 'a workshop for production' (Scott 1998: 115). But the city as it is lived is a messier, inefficient, conflictual state of affairs, and thus governors feel compelled to gloss over that reality: 'A great deal of the symbolic work of official power is precisely to obscure the confusion, disorder, spontaneity, error, and improvisation of political power as it is in fact exercised, beneath a billiard-ball-smooth surface of order, deliberation, rationality, and control' (Scott 2012: 140). As Isin argues, 'we must be skeptical about claims made by citizens about the harmony, unity, and homogeneity of the polity they claim to represent' (Isin 2002: 29), and look further at the power relations and conflicts that exist below the surface, in the city as it is actually experienced.

It is within this contested landscape with its façade of harmony that citizenship is negotiated. While a common understanding of citizenship is its formal status with its attendant rights to vote and carry a passport, this is only the most visible and tangible articulation of a more complex institution (Isin and Wood 1999). Citizenship is not merely a status or institution, but a relationship between the governed and the governor – which may take other forms and involve other parties than the State (P. Wood 2014; Isin and Ruppert 2015). Although the specifics have changed dramatically over time and place, Rancière reminds us that Aristotle 'defines the citizen as 'he who partakes in the fact of ruling and the fact of being ruled"' (Rancière 2010: 27). It is a technology of governance on the part of the State, to discipline and define its constituents. It is also a mechanism for making claims on the State by those considered to be constituents and for those who constitute themselves as political subjects deserving of a claim – even if they are not formally recognized as such by the State in the first instance. Formal status is neither a guarantee of one's rights and freedoms by the State, nor necessary to make successful claims (P. Wood 2014; Smith 2001).

Citizenship also explicitly claims equality: it is a political status alleged to transcend class, race, gender, ethnicity, religion, sexuality, physical ability, and so on. Yet it is deeply and not coincidentally embedded in the creation, definition and manipulation of those categories of identity. Like the myth of the unified city, through its interrelated narratives of unity, equality and the 'good citizen' (White 2006; Wood 2001), citizenship *disciplines.* Definitions of the citizen have been explicitly predicated on identity categories, as it restricted (for example) voting and property rights only to certain groups and even excluded certain groups from holding citizenship at all. These occur in formal, legal practices (such as an earlier exclusion of Asian people born in the United States from holding citizenship or the inclusion of 'non-Germans' born in Germany) and in informal social practices that stipulate modes of national identity specific to particular bodies.

Many of these excluded individuals and their lives were previously (and in many cases remain) invisible or expendable to organized power, to capital, government and organized labour. They are not powerbrokers or lobbyists. Their work was at the edges of decisions, often struggling in the wake of those decisions: struggling to make rent or pay the mortgage, scrambling to find somewhere safe to live, teaching in a classroom with hungry children, evading the border police, discussing the latest budget cuts with other disabled friends. Their politics often did not find resonance with party platforms because their politics were often 'impossible'. What they – *we* – want is impossible: the transformation or reversal of existing social and economic orders, for things to mean other than what they currently mean.

The political philosophies and subsequent strategies of communities whose identities are 'other' are 'impossible'. The neoliberal city empowers the wealthy, gates its residences, and negotiates its governance through private networks of corporate boardrooms and golf clubs, as much as through legislative bodies. These are all spaces where poorer, less privileged people *work*, where their entry is through the backdoor, after hours, behind closed kitchen doors, moving in and out of rooms to clean, bring coffee and take notes, without speaking (Isin 2002; Sassen 1997). The patriarchal city empowers men, not women, and they occupy different places and spaces in its landscape. The racist city empowers whites and whiteness, and the processes of categorization are also spatial and concrete in their allocation of privilege, of mobility rights, of political voice. Women, racialized persons, gay, queer and trans persons cannot move through the city in the same way – not only because they are rarely in the limos or even the taxis, but because their presence on the street is formally and informally policed. They are more frequently stopped by the police, without cause. They are subject to street harassment. They are vulnerable to attack on the streets at night, by white supremacists, homophobic thugs, rapists, even by the police. Where can they organize? Around what issue or identity? Is theirs the kind of reality where a political strategy of setting a clear agenda to bring to the legislators or the courts makes any sense? Or is likely to succeed? Maybe. But not on its own. Nor would it be a sufficient answer to the violence they face.

I would like to support but expand Gerbaudo's argument that we need to approach activism in its specificity, in terms of the embodiment and emplacement of protest. Despite seeking a balance in terms of gender, race and so on in his interviews, there is nothing in Gerbaudo's text on the role of gender or race/racialization in activist practices. He cites Hardt and Negri (Gerbaudo 2012: 27) to the effect that the differences among diverse participants are immediately 'transcended' by the collective. But we may ask, in what way is the 'multitude' of Hardt and Negri (2004) 'varied', except superficially or aesthetically? Where is the difference those different embodiments make, politically, in philosophy or practice? What difference does the antagonism among them make? We will return to the larger question of democracy and the 'multitude' or the 'people' in the first and fourth chapters. However, by looking more closely at some of the communication technologies Gerbaudo discusses, we can immediately see more evidence not only of uneven participation and access, but of differentiated treatment. What Castells misses in his enthusiasm for the Internet is how Twitter users are vulnerable to state surveillance, and also to trolling, stalking, doxxing and other forms of interpersonal violence, that have at times escalated offline. These forms of violence are directed overwhelmingly at women, women of colour, queer persons, and members of the trans community. Just because 'everyone' can speak to each other does not mean they listen and respect each other (as Twitter surely also demonstrates). As Bissonette notes:

> One of the most seductive aspects of the digital age is its potential to create a virtual democratic model of participatory cyber-citizenship. [However,]...the anonymity of participatory media has become a harbinger of an uncivil society. [And]...the forums where a new breed of citizen journalists could freely debate current events have become caught in a struggle between digital democracy and twenty-first-century tyranny.
>
> (Bissonette 2014: 385)

What many hope for and thus see in Tahrir Square or Occupy is a new solidarity, to overcome a 'condition of social individualisation' that Gerbaudo discusses (2012: 30), also identified by many theorists (e.g. Hardt and Negri, Baumann, Beck, Berardi) as a new fragmentation of previously common, solidaristic identities that were seen as the basis for collective action. But the claim of individualization is misleading. It has not happened to everyone, and the previous alleged solidarities are over-stated or even non-existent. Such aspirational solidarities run parallel to the case of the middle class who can comfortably adopt national identities revolving primarily around class, and step away from other 'divisive' and 'tribal' identities. Not everyone can do that; to do so has always been predicated on denying the specifics of those 'previous' identities, losing any solidarity with them. A fundamental assumption for citizenship, for Weber, was that residents of Western cities let go of 'tribal' identities and transcend them in the city (in a new 'fraternal association'), but this is not the reality (Isin 2002: 11). This is the city that the dominant groups, those who claim and purport to define 'citizens', have invented. In the 1960s, for example, Aboriginal peoples in Canada were offered suffrage literally in exchange for giving up their 'Indian status' and any particular rights and privileges associated with it. The expected exchange is plain: the 'citizen' is supposed to separate himself (always *himself*) from 'narrower' interests. The resulting premise is that Western citizenship constitutes the antithesis of such tribal attachments, and creates the unified constituency who logically inhabits the unified city.

This book is about the corners of the Western world where the promise of Western citizenship has yet to arrive. It seeks, as Cathy Cohen's work does, the articulation of a politics for those whose identity is other than 'state-sanctioned, normalized White, middle- and upper-class, male heterosexuality' (Cohen 2004). Cohen (whose particular research focus is African-American youth) asks how these people, 'who often *rightly* perceive themselves as tangential to American democracy, create and understand their political lives?' (Cohen 2010: 16). The embodied experience for many women, Black men and women, women and men of colour, gays and lesbians, queer persons, indigenous peoples, people with non-dominant residential practices, and disabled people, is a struggle. The 'citizenship gap' between constitutional statements of equality and lived realities of excluded and exploited bodies is stark, and is neither incidental nor accidental (Berlant 1997; Wood and Wortley 2010). Through citizenship, the nation articulates its desired, worthy bodies, and demands that other bodies mimic them. As Patricia Williams wrote of blackness:

> The blackness of black people in this society has always represented the blemish, the uncleanliness, the barrier separating individual and society. Castration from blackness becomes the initiatory tunnel, the portal through which black people must pass if they are not to fall on their faces in the presence of society, paternity, and hierarchy.
>
> (Williams 1991: 198)

Our everyday lives are politicized into protest and resistance, sometimes without any decision on our part. Fundamentally, these lives should invite us to reconsider any assumption that being political is an alternate state, separate from the everyday (Cohen 2004; Chatterton and Pickerill 2010; Runciman 2016; Véron 2016). The dividing lines between everyday life, community activism and mass protest are blurred. These connections also reframe our assertions or critique of the 'spontaneity' of protests. Everyday life, individual experiences and daily sharing of information may politicize and mobilize, on a variety of scales and issues, which may then spiral into larger campaigns or connect to related issues. Or they may not.

I argue for a continuum of acts of resistance to suffering and exclusion, both private and public, that includes the less visible, less spectacular, in addition to mass mobilizations. Drawing on the ideas of Cathy Cohen, Audrey Wollen, Joanna Hedva, Davina Cooper and Lauren Berlant, among others, this continuum includes behaviour labelled 'deviant', non-compliant, uncooperative, queer, creative, and outrageous. Both the quieter daily negotiations and the explosions of 'diva citizenship' (Berlant 1997) transform the space of the city. This is resistance and citizenship, and its spatialities are critical to identity and belonging. Moreover, individuals and communities work at claiming, reclaiming, and repurposing space to create an environment in which to survive and thrive. What these invisible people are doing is remaking the city, and in their acts, they are living their politics, grounded in the realities of their lives and bodies. They are creating the space in which alternatives can happen, reinventing themselves as citizens. Specific acts are often temporary and ephemeral, but the work is constant.

The assembly of ideas here subscribes to what Sedgwick (1997) and Gibson-Graham (2006) term 'weak theory', as an alternative to the totalizing master plan of 'strong theory'. 'While [strong theory] affords the pleasures of recognition, of capture, of intellectually subduing that one last thing, it offers no relief or exit to a place beyond', and Sedgwick has argued it is a form of paranoia (Gibson-Graham 2006: 4). Cooper (2014) similarly challenges the idea of a singular theory that completely and permanently frames our investigations. In its place, they recommend we allow for uncertainty, change and unpredictability. In many ways, this starts with the archive (see Berlant 1997; Smith 2002) and choir of theorists upon which one rests the research. I have been studying and working with communities whose ways of living, articulations of identity, and understandings of history do not *fit* with normative definitions of citizenship. Often their very bodies are deemed impossibly imperfect or dangerous for inclusion as 'citizens' (Yingling 1997; Lorde 1984; Hedva 2015; Brown 1995; Lister 1998; Knadler 2013; Breckenridge and Volger 2001; Crowley 2005; Simpson 2014). As the poet Dominique Christina says, 'I am a black girl. I interrupt space everywhere I go' (Emmanuele 2015). Their realities are not accepted by others, particularly in the mainstream. Their stories of unfairness, dispossession, and assault are challenged or disbelieved, their motivations questioned. Their systems for producing, articulating and sharing knowledge are dismissed or actively delegitimized. Their emotions are similarly dismissed or worse, used to discredit their testimonies. These experiences are my archive. There is no one master theory that explains the multiplicity of power relationships at work, although intersectionality (Crenshaw 1989) is a fundamental aspect. It is similarly of limited value to start with mainstream sources and theorists and/or to measure things quantitatively to gain a better understanding of the perspective and politics of individual communities ignored by politicians, mainstream media and, often, scholars. These processes of marginalizing or silencing alternate perspectives are also present within the academy (Klodawsky et al. 2016), and diversifying the archive and theoretical choir of research to include marginalized voices, to de-colonize and otherwise de-toxify our methodologies (Smith 2002) is a necessary step.

Even after more than two decades at it, I cannot claim to have checked all my privilege or to have fully decolonized my research. I can only claim to have made an active effort, to have prioritized listening in my learning, and to work towards amplifying voices whose suffering has been marginalized and discredited. Here, I have drawn actively on insights from (none of these is mutually exclusive of another) feminist, queer, Black, anti-racist, indigenous, critical disability and other non-dominant/non-normative theory. These writings present a variety of forms, perspectives and starting places. Among their other insights, these bodies of theory take different form and produce different types of knowledge:

> For people of color have always theorized – but in forms quite different from the Western form of abstract logic. And I am inclined to say that our theorizing…is often in narrative forms, in the stories we create, in riddles and proverbs, in the play with language, because dynamic rather than fixed ideas seem more to our liking.
>
> (Christian 1987: 68; see also King 2003)

These ideas are part of the intellectual but also the social life of marginalized communities, and Christian argues the joy and playfulness of language and action that are present in their politics have been critical to surviving the constant battering they experience (Christian 1987).

In addition to my own specific fieldwork over the years, I want to note that I have found social media, particularly Twitter, to be an excellent place to consciously expose oneself to a wide variety of voices, and just sit back and listen. My goal here, as ever, is to take the so-called 'margins' as my starting place (Wood 2001), from which to walk a different and hopefully insightful path through the critical concept of urban citizenship. My methodology is grounded in beliefs that our knowledge and epistemologies are determined in the first instance by the limits of our epistemic frame (of which we are not aware without critical reflection), that with critique it is possible to learn and expand one's understanding of what constitutes reality and knowing, and that new, alternative possibilities are often best expressed in un-measurable and even artistic and fictional forms.

To incorporate these lives into dominant theories of the city and politics requires opening our understandings of oppression and politics of resistance much more broadly, allowing for a much wider range of actions and of how they connect and synergize. This would entail a politics that fundamentally theorizes oppression and resistance, without creating a hierarchy of the type, scale or location of either. It must be informed by feminism, anti-racism, anti-homophobia, anti-colonialism, anti-transphobia, anti-queerphobia and critical disability studies. This theorizing is thus grounded in and informed by lived experience, although we must also continue to incorporate larger political economic theory about the way the city is formed and how social relationships are mediated by capital and markets. If we seek to relieve and end suffering, we must recognize it in every form, with equal urgency. Building on refusal politics, diva citizenship and sick woman theory, I also consider anarchist critiques, particularly Emma Goldman's, of what politics itself is, where and for whom, as a path to understanding the political acts and social movements of the present. For Goldman, the barbershop, the salon, the doughnut shop, the playground, the local library, and the pub were important political spaces. I argue for seeing human beings as political subjects, as always citizens in the making, negotiating their desires for security, autonomy, joy, and productivity with all parties who seek to govern those choices, and in a way that does not pathologize difference or deviance. This would ideally also be grounded in a theory of justice that recognizes all forms of oppression, suffering, trauma and injustice.

Mass mobilizations are an important part, but only one part, of a larger set of practices of urban activism and protest in the early twenty-first century. Occupy was bigger than the encampment at Wall Street, and smaller than post-crisis political activism – not just because it was global, but because Occupy was and is just one highly visible articulation of resistance to capitalism, gentrification, austerity, privatization, surveillance, and so on. But those movements and even the term, 'Occupy', came to stand for all those politics of resistance, and the closure of a camp a sign of its failure.

This book is an effort to develop further a radical urbanist's framework for understanding activism and protest, the work it does or does not do, and what it means for the current state/future of politics and the city. I want to reorient our understanding of post-crisis activism away from Occupy – not to forget or dismiss Occupy but to contextualize it differently and critique the privileged lens many have applied to it. I also seek to rethink the motivations and affects of activism and protest, to (re)ground them in suffering and the emotional, not purely the rational, collective and analytical. This necessitates a return to the idea that the personal is also political – politics is not only in personal issues, but in private spaces, and those spaces, including the body, are sites of political struggle. Personal encounters with white supremacy, patriarchy, homophobia are the result of a larger societal sickness that is part of society's structure and normal functioning. Occupying and having some control over those spaces are regularly contested.

There are several things I hope these essays will do. I want to unpack current discourses of mass occupations and explore our expectations of citizenship, activism and politics, and the construction of their legitimacy and success/failure. I also seek to re-examine the city, to see it as an inhabited city, to grant its social geography some agency in the shape of the city and some autonomy from the city's economic geographies, to see the alternatives that already exist. Ultimately, this will enable us to rethink what we mean by politics, by social and political movements, and how we measure the success and failure of our actions. With the help of a discussion occurring largely in Italian and French scholarship, I want to argue that the idea of 'democracy' supported by 'the people' is a myth that can never be embodied, but one that can be institutionalized through the constituent 'documents' of citizenship. With some further assistance from Andreas Kalyvas' (2008) ideas of constituent power and 'extraordinary politics', I want to argue, in conclusion, that these politics of the invisible, with their invisible resistance and impossible demands, are in many instances a higher order of politics. Mobilizing politically – that is, enacting citizenship – within such political and social landscapes, has its own particular logic. There are some social movements that seek specific legislative change or political office. But much activism does not. Much, possibly most, activism seeks to work at a local level, to improve the material and discursive culture of the everyday, to resist violence, to alleviate suffering. Activism and protest, in all its forms, is far from a lower order of politics; rather, it is often a kind of constitutionalism that is a powerful act of citizenship.

In some ways, this book is a postcolonial sequel to the political economy approach, the *Wide Sargasso Sea* to its *Jane Eyre*. These are the politics of the madwoman from the colonies imprisoned in the attic (see Gilbert and Gubar 1979). With insight from the humanities on language and art, the political and social sciences can restore the messy embodiments and emotions they have stripped out of their studies. Our emotional attachments to place – love *and* anger – are what make a different politics possible. What I argue for here is a framing of urban politics that is grounded in the concrete, in the embodied experience of inhabiting the city, and a recognition that these politics are constituted differently, and in a diversity of ways. They are no less legitimate or effective than public, institutional and formal politics. A broader critical understanding of oppression, of harm, of trauma, of injustice, and of the many forms of emancipation from them, must inform our understandings of the working of the city as actual lives intersect with, navigate, and sometimes even avoid the capitalist political economy of the city. The politics of our own politics, the way we privilege certain acts and see in those acts what we want to see, repeats the exclusions that have oppressed and traumatized others. The ability of the marginalized, the exploited and the traumatized to explode the frame of discussion through line-crossing, even outrageous acts may produce new possibilities in the space blown open, before the old ideas fall back into their comfortable places. It may be ephemeral, even invisible, but it is a transformation of space, and through that spatial rupture is born a moment of emancipation, belonging, truth, and hope. That hope and that belonging are essential to act as a citizen, and they speak to nothing less than our principles of government.

Note

1 Reproduced by permission of Penguin Books Ltd.

1 What we talk about when we talk about Occupy

Politics and citizenship in crisis

This chapter is not about Occupy New York. It isn't really even about Occupy as a movement at all. What I want to explore here is how 'Occupy' has entered the vocabulary and discourse of academics, activists and the general public to produce and reproduce our expectations of activism, protest, democracy and citizenship. 'Occupy' remains a placeholder for the discussion of protest in the wake of the crisis, used by both fans and critics as a measurement of politics to be embraced or avoided. It has found a place in our vocabulary in much the way ''68' did and still does, and is as densely coded with a variety of meanings. In the case of the Occupy movement itself, the field is distorted by a US focus and muddied by a multitude of understandings of what 'Occupy' even is or was. A strong confirmation bias, particularly regarding the rise of a 'new Left', also exists in the literature, where various scholars largely see what they want to find rather than approaching Occupy on its own terms, in its specific contexts. With the inclusion of voices and perspectives from outside the US and outside the English-speaking world, I have tried to counterbalance and enrich these understandings through an analysis of the discourse that pays more attention to the work the signifier 'Occupy' has done during and since the occupations, as well as to the limitations and blind spots of that discourse, rather than an assessment of the validity of its content. I include in 'discourse' not only textual and visual representations and their connotations, but also the ideas of what Occupy 'is' and what it achieved or served through its presence and practices. In other words, I take no position on whether or not Occupy is the rebirth of '1968', or if its horizontalism is its greatest strength or a fatal weakness. I am interested in what a close look at the discourse of Occupy tells us about expectations of citizenship.

If we accept that every occupation of any kind that called itself 'Occupy' is part of a single movement, and if we accept that Occupy is in fact a movement, then Occupy is one of the most widespread spontaneous social movements the world has seen. For many, it is also one of the most short-lived. If there is anything that is as striking as the unexpected success of Occupy's diffusion – to over 1500 occupations in approximately 100 countries, it is the apparent failure of Occupy to achieve what its participants and observers expected of it.

'Occupy' gave everyone – scholars, politicians, union leaders, the media and the general public – a descriptive handle to grasp, a vocabulary of 'occupation' and the '99 per cent'. But that vocabulary did not capture the deeper history, politics and perspectives of the protesters. When we talk about Occupy, either with specific reference to 'what happened' and 'what it means' or as a metaphor for other post-crisis politics, we are doing many things. Pulling apart 'what we talk about when we talk about Occupy', however, goes beyond the critique that 'everyone sees what they want to see' – although that is one important component. Occupy is a signifier that codes politics, narrative, and aesthetics, as well as the literal and discursive space of the city.

Rather than taking Occupy as an exemplar, or even simply an object of study in and of itself, I examine discourses around Occupy as a path to understanding the politics of protest and citizenship in the early twenty-first-century city. I have not conducted a systematic review of every mention of the word 'Occupy' in the newsmedia or on the Internet, but I have spent several years collecting references to it, both first impressions and later thoughts, from around the world, particularly when Occupy was being used to talk about something else. I combine a review of the discourse in scholarship on Occupy with my own analysis of what was at work and put to work in 'Occupy' with these references, and find a diversity of meanings, usages and connotations. Ideas about Occupy and the mass mobilization it appears to exemplify both reflect and structure ideas about politics – not just with respect to 'appropriate' or 'effective' ways of protesting, but also about fundamental ideas of democracy. Reflecting on the different types of discourse at work, rather than the different political positions, I have sorted these into four categories: politics, narrative, art and grammar. These categories are not simply pots in which to gather like things, but angles from which to consider the work that Occupy did and continues to do – what it emerged from, how it has been understood, what it reinforced and what it changed. By examining the positive and problematic aspects of this discourse, as well as what and who is missing, I also consider how limited Occupy was and remains in its impact, and in its capacity to 'occupy' a city and represent protest.

I argue that what is commonly thought of among sympathetic media and academics as authentic, real, or deep democracy is more problematic than it first appears. Its particular spectacle of mass mobilization is often what some think activism is and should be. But its 'grammatical' invocation of the people, and related ideas of democracy, is narrower that its presentation. I argue that Occupy is misrecognized first, as the 'start' of a new beginning that will be the rebirth of the Left, democracy and/or ourselves, possibly even through the formation of a 'global social body' (Hardt and Negri 2004); second, as a more authentic and real form of politics, for which we have been waiting; and third, as an inclusive, solidaristic mobilization that represents the 99% or 'the people'. This last idea will entail a more detailed exploration of how the idea of the 'people' functions vis-à-vis democracy, and this discussion will be taken up again in the final chapter. I am not saying that Occupy is none of the things people believe it is; I am observing how deep and wide these ideas run, and arguing that they are more complicated than they first appear.

Occupy as politics

The most obvious and dominant use of 'Occupy' as a signifier is with regard to politics. For many, especially on the Left, the Occupy movement signals a type of politics, but there is a variety of meanings as to what that politics is. For some it is new way of politics and the way forward; others see it as the culmination of existing political movements and strategies, and 'real' in a way that even other mass protests are not; still others, on both Left and Right, see it as representative of a type of politics that does not work.

Occupy Wall Street began 17 September 2011. Slavoj Žižek came and gave a talk on 9 October. 'We are the awakening!' he said. Žižek also told the occupiers, 'Don't fall in love with yourselves', advising them that they needed leaders, structure, a plan (Žižek 2011). On 11 October, Noam Chomsky gave a talk at Occupy Boston, as part of a named lecture series for Howard Zinn. These visits to a leaderless movement from leading activist academic figures were a particular kind of performance that signalled the importance of the occupation: this was 'real' politics. Their presence announced it as such, even before their words confirmed it.

Occupy has received a certain glorified attention, and many first impressions described a historic moment of the emergence of a new politics that had been a long time coming. It was not merely an event in and of itself but the beginning of something larger in terms of political change. Authors of a series of short papers on Occupy Wall Street (OWS) written in its immediate aftermath for the journal *Contexts*, were generally effusive:

> 'dissidence of a sort that we have not seen in this country for a very long time'
> 'like New England town meetings with an anarchist edge'
>
> (Contexts editors 2012: 13)

> '"horizontalism," a form of consensus-based participatory democracy that is inclusive and anti-hierarchical, as well as a "prefigurative politics" that aims to foreshadow a democratic social practice inside the movement itself'
> 'destined to have a lasting impact on American politics and culture'
>
> (Milkman 2012: 14)

> 'a revelation'
> 'a bold attempt to embody an alternative paradigm of participatory engagement'
> 'do not have demands, but they do have an agenda.'
>
> (Barber 2012: 14–15)

> 'the Occupy movement is utopian and practical – a better world can be created, not in the distant future, but right now.'
> 'Someone ought to thank Occupy for accomplishing in a few short months what sociologists have been unable to achieve over decades'
> (Williams 2012: 20)

> 'how motivating it can be to engage in collective self-governance and develop new social relations, to come to know your own and other's intelligences and capacities, and to be changed while building new worlds'
> (Gould 2012: 21)

Noam Chomsky called it 'the first major public response, in fact, to about thirty years of a really quite bitter class war that has led to social, economic and political arrangements in which the system of democracy has been shredded' (Chomsky 2012: 54). Rebecca Solnit includes Occupy as one of those

> mysterious moments when citizens felt impelled to act and acted together, becoming in the process that mystical body *civil society*, the colossus who writes history with her feet and crumples governments with her bare hands....[T]he old order is shattered, governments and elites tremble, and in the rupture civil society is reborn.
> (Solnit 2013: ix)

She more than hints at the rebirth of a nation: 'In these moments of rupture, people find themselves members of a 'we' that did not exist, at least not as an entity with agency and identity and potency, until that moment' (x). For Nathan Schneider, 'Occupy Wall Street was such a momentous thing and such a rare moment of political hope for us who were born during the past thirty years in the United States of America' (Schneider 2013: 5). He refers to it as an apocalypse, in the literal sense of 'the lifting of a veil' (6). Micah White said, 'This is a moment for all of America....[T]his is a moment for us to reinvent democracy in America...' (Schneider 2013: 15).

Greg Ruggiero describes Occupy, in its philosophy and pragmatics, as something more than mere politics, as a 'vision':

> In no rush to produce leaders or to issue a closed set of demands, Occupy embodies a vision of democracy that is fundamentally antagonistic to the management of society as a corporate controlled space that funds a political system to serve the wealthy, ignore the poor and answer everyone else the same way the politicians answered Anderson Cooper: however the hell it wants.
> (Ruggiero 2012: 15–16)

David Graeber, while a professor of anthropology at the London School for Economics, was involved with OWS from its beginning and is credited at least partly with coining 'We are the 99%.' Graeber considers Occupy to be an enormous success. As Solnit also did, Graeber directly compared OWS with 1968 – and 1848 (Graeber 2013). Castells also compared what he saw as the emergence of 'a new pattern of social movements', even prior to Occupy, to his 'personal experience, as a veteran of the May 1968 movement in Paris'. Interestingly, Castells immediately describes this comparison as a personal, affective one: 'I felt the same kind of exhilaration' at the possibility that this might be the dawn of 'a new revolutionary era, an age of revolutions aimed at exploring the meaning of life rather than seizing the state' (Castells 2012: x). Some argued that with its inclusive framing, Occupy brought a new and peerless inclusivity and solidarity: 'It was a beautiful movement, because the definition of 'we' as the 99 percent was so much more inclusive than almost anything before, be it movements focused on race or gender…' (Solnit 2013: xi). Milkman similarly suggested the participants at OWS were 'a more racially and ethnically diverse group than is often presumed' (Milkman 2012: 13). Graeber also argued that Occupy's alliances and solidarities were new and encouraging (Graeber 2013).

In a review of publications immediately following OWS, Alisdair Roberts observed, 'They are energized by the spirit of the moment. Unfortunately, they are also blinded by it. The prevailing theme of these books [Byrne 2012; Gessen et al. 2011; Gitlin 2012; van Gelder 2011; Writers for the 99% 2012] is that OWS has fundamentally changed American politics' (Roberts 2012, 754). In October 2014, Michael Moore tweeted that 'Occupy Wall Street changed the ENTIRE conversation'. Harcourt argued it was a new type of politics:

> In its steadfast refusal to compromise with political power, to conform to conventional politics, or to play by the rules, Occupy Wall Street immediately fashioned a new form of political engagement, a new kind of politics.…To turn a phrase, I would say that Occupy Wall Street instantiated a new form of 'political disobedience' – a type of *political* as opposed to *civil* disobedience that fundamentally rejects the ideological landscape that has dominated our collective imagination, in the United States at least, since before the Cold War.…Occupy Wall Street is politically disobedient to the core – it even resists attempts to be categorized politically. The Occupy movement, in sum, confounds our traditional understandings and predictable political categories.
>
> (Harcourt 2013: 46–47)

Harcourt argues that Occupy is part of 'an emerging fourth American left' (Harcourt 2013: 63). Not for Harcourt alone, OWS is a sign of the long-awaited rise of the 'new Left', the rebirth of activism and a popular confrontation with neoliberalism. These commenters have a particular focus on Occupy New York, and the United States, and the power that is Wall Street in particular.

For others, 'Occupy' as politics specifically refers to its deliberative practices of 'deep democracy' and prefigurative organization. As Barber wrote, 'To be authentic, democracy must be palpable, participatory and local – like Occupy

Wall Street itself' (Barber 2012: 15). Castells observed a turn towards deliberative practices as part of a turn away from traditional political institutions and practices: 'In all cases [of Occupy, worldwide], the movements ignored political parties, distrusted the media, did not recognize any leadership and rejected all formal organization, relying on the Internet and local assemblies for collective debate and decision-making' (Castells 2012: 4). They held highly participatory group meetings, with use of the 'people's microphone', which Taussig argues was 'the most powerful invention of the Occupy movement', so that everyone can hear and know what's going on (Taussig 2013: 21). For what became OWS, the planners went with 'Keeping tactics loose'. 'The process of bottom-up direct democracy would be the occupation's chief message at first, not some call for legislation to be passed from on high. They'd figure out the rest from there' (Schneider 2013: 20). He notes that some of this arose from suspecting there were cops at the general assembly meetings, so too much specific planning ahead made them vulnerable to being thwarted. 'Maybe assemblies like this could even become a new basis for organizing political power on a larger scale' (21). In this new kind of politics, there remained the implicit idea of larger scale as greater success.

Noam Chomsky saw in Occupy what we might call a pilot test of another world: in the miniaturized economy, the collective, communal kitchens and libraries, he saw proof that alternatives to market relations are possible, even easily possible. But he moved quickly from prefigurative politics to aspirations for mass disruption modelled on the labour movement:

> And I think that maybe, in many ways, the most exciting aspect of the Occupy movement is the construction of the associations, bonds, linkages and networks that are taking place all over – whether it's a collaborative kitchen or something else. And, out of that, if it can be sustained and expanded to a large part of the population who doesn't yet know what is going on. If that can happen, then you can raise questions about tactics like a general strike that could very well at some point be appropriate.
>
> (Chomsky 2012: 45)

Another aspect of its significance was as a link in the chain of a larger, global politics: 'Occupy is, as I've repeatedly stressed, simply the North American manifestation of a democratic rebellion that began in Tunisia in January 2011, and by the end of that year was threatening to call into question existing structures of power everywhere' (Graeber 2013: 130). Purcell connected the 'Arab Uprisings in Tunisia, Egypt, Libya, Syria, Yemen, Bahrain, and Oman in 2010–11' and speculated that they were 'some of the most spectacular manifestations of raw constituent power in recent memory' (Purcell 2013: 14–15). In light of these mobilizations, Purcell argued that 'Democracy, the democracy in which people struggle together to take control again over their own lives, is back on the agenda', and that this was a global struggle (29). Bamyeh agreed that 'a new global culture of protest began to take shape in 2011' (Bamyeh

2012: 16–18). For Chomsky, these connections were an opportunity to stress the potential role for the labour movement:

> In fact, one striking difference between the Egyptian and Tunisian uprisings and the Occupy movements is that in the North African case, the labor movement was right at the center of it....As soon as the labor movement became integrated into the April 6 movement – the Tahrir Square movement – it became a really significant and powerful force. That's quite different here. The labor movement has been decimated. Part of the task to be carried out is to revitalize it.
>
> (Chomsky 2012: 58–59)

Occupy has also been seen as a failed politics: a useless politics, a protest that lacks structure, strategy and an agenda, and thus is not actually politics. Although Murray Bookchin had passed away before Occupy launched, he had previously written with reference to the protests against the World Trade Organization meeting in Seattle that, 'A politics of protest is not a politics at all' (Bookchin 2015: 171), seeing them largely as the product of 'lifestyle anarchism' that had no theory to underpin and guide it. In May 2012, Andy Ostroy wrote in the *Huffington Post*, 'The Occupy "movement" (and I use that term generously) has spiraled into irrelevance and relative obscurity.' This was due, he argued, to a focus on actual occupation which 'confined and defined the message in a way that was limited and negative', had 'no leadership', 'no agenda', and 'the wrong message' (he felt it should have been 'fairness for all through reasonable government regulation and taxation') (Ostroy 2012). For Harcourt, Occupy was unsuccessful, not due to its own weaknesses but because of the forces that shut it down: 'the peaceful voices of Occupy and other protests were simply drowned out. They were buried beneath the hype and frenzy of our police state, the legions of black-clad riot police with batons, the images of a few violent clashes, and the frightening images of our new normal urban militarized security' (Harcourt 2013: 81).

Christian Parenti critiqued the movement for its failure to engage formal institutional politics:

> the state cannot be avoided, as scholars like Holloway suggest. For Left politics to be effective movements, especially in the face of the climate crisis, they must come up with strategies that engage and attempt to transform the state. The idea of escaping the state is to misrecognize the centrality and immutably fundamental nature of the state to the value form and thus to capitalist society.
>
> (Parenti 2015: 843)

Parenti argues that we cannot bring about radical change without the State, particularly when it comes to large-scale, global problems such as climate change. Occupy is thus a kind of large-scale politics operating in the wrong sphere.

As a mode of politics, Occupy often was and remains in the eye of the beholder, particularly as it is made to fit into pre-existing conversations on the Left, with limited expectations of its specifics. Those more directly involved, especially Graeber, have greater knowledge and a more nuanced appreciation, but their vested interests may shape their impressions and the ways in which they continue to represent the event.

Occupy as a story

Social scientists need to take stories more seriously. In the humanities, it is easy to find arguments that stories are all we are (King 2003). Neuroscientists are also starting to confirm these findings in the humanities by documenting that stories are how we learn, that it is through metaphor – constant comparison and connection – that we build and process knowledge (Lakoff and Johnson 2003; Ramachandran 2011). Our brains respond to stories by activating the same parts that 'light up' when we are experiencing something physically. Stories are an art form, but they are more particularly an attempt to understand our realities and to create 'truth'. Thus, the stories and metaphors mobilized by a discourse become lessons, in which the pattern, scope, and meaning of things are *explained*, and are thus powerful messages.

Occupy is a story we tell, to ourselves and to each other. It is at times a specific story of 'what actually happened', but it is just as often a story of the entire global financial crisis and public response, of why it happened, of what is wrong with the system (or not), and what can and cannot be changed. When we talk about Occupy, we talk about what we think is possible and impossible, what it means to be political, what that is supposed to look like, and who are legitimate political actors (who speaks and who just makes noise, as Rancière would say) even amongst members of the broad activist community. Occupy did not invent any of this, but the scale of Occupy brought it to the fore, made 'Occupy' a vehicle and vocabulary many would use to understand the crisis and our politics, and therefore a reference point for other political movements. Narratives of Occupy fall into long-existing categories already common in Western storytelling. Of the many stories people tell through 'Occupy', I want to highlight briefly the emergence of three: the *conte morale* or lesson for other activists, the tale of rebirth, and the heroic epic.

When the Ferguson and Black Lives Matter (BLM) movements started in 2014, hopes were immediately expressed that the movement should learn from Occupy and not fade away like a passing trend. A column in *The Root* argued that Occupy 'fizzled out because it was unfocused and disorganized and lacked leadership. The nation can't afford for the Black Lives Matter movement to be nothing more than a racialized version of Occupy Wall Street, wherein anger was given voice, but change didn't materialize' (Johnson 2014). Former New Zealand prime minister Helen Clark expressed the same fear: 'We saw Occupy flare up and then fade away like many others like it', she said. 'The problem movements like these have is stickability. The challenge is for them to build structures that are ongoing; to sustain these new voices' (Hogg 2015). Like Žižek's criticism to OWS to get organized and not just play around, movements that came after Occupy often recounted it as a story of a failure to organize.

Here, Occupy was a cautionary tale of mobilization-without-staying-power. This is a story of directionless actors, with strong emotion but incomplete character development. The moral of the story for activists: know who you are and what you want; movements have to have an agenda, goals, and a structure to sustain them. Occupy is a lesson about the timescale of politics as well: it requires staying power, because the road is long. The victory of change is not the result of any one moment or one act or one decision. But the cautionary tale is also an effort to discipline activists into certain kinds of behaviour and certain kinds of political goals. Even among those whose political goals were similar, there was a story of the correct way to be an activist citizen, an argument for being political in particular ways.

Among some participants and observers, Occupy was frequently a story of rebirth, an enlightenment or awakening: 'This momentous historical period of uprisings around the globe reflects…the political engagement and awakening for hundreds of thousands of first-time activists' (Ratto and Boler 2014: 24). A metaphor of awakening or enlightenment is not unique to Occupy; for example, BLM highlights an idea of becoming 'woke'. The sense of new ideas, new participants and a new beginning was individual and collective. Sometimes the participation of many young people in Occupy was held up as a counter to accusations of their political apathy and the renewal of democracy. Individuals achieved new awareness, but it was also communities and nations that were reborn or revived in the process.

In a speech on 15 October 2011 in Dublin, Helen Sheehan described the new growth that was Occupy Dame Street:

> These occupations have opened new spaces, literally as well as metaphorically. They have created a new dynamic in this struggle. They have transformed the terrain, physically here in Dame Street, but psychologically and politically as well.
>
> How? Well, first of all, we've transformed ourselves. We have become the change we want to see. Those who have stepped up for the first time can never be the same again. You have discovered something new in yourselves.
>
> (Sheehan 2011)

Sheehan speaks of a transformation process that is both psychological and political. Miriam Sitrin's account similarly suggests fundamental personal and social change, 'Occupy changed everything. The U.S. is not the same place – there was a before and after.' Following the experience of Occupy, there were not merely lessons learned, but 'new ways of doing things, new ways of being' (Sitrin 2014). The implied cycles of birth/death, sleep/wake were collective and connected, where rebirth and renewal were interactive, caused and maintained in/by relationships with others.

Another story of Occupy and other occupations revolves around its conflicts with authorities, as a narrative of a kind of heroic politics. In his discussion of the politics of 'gray spaces', communities which are marginalized yet not fully excluded or eliminated, Oren Yiftachel notes that 'the struggle rarely entails heroic confrontations with the authorities, nor does it produce comprehensive strategies or finely defined agendas' (Yiftachel 2011: 152). His comment is not a criticism of such politics, but the particular language he uses to describe this observation flags an important aspect of the stories we tell about protest. For many groups mobilizing against the injustice of the status quo, one goal is a form of direct action involving 'a dramatic confrontation with armed representatives of the state' (Graeber 2011: 64). Even in the absence of an individual *hero*, certain types of political action are viewed – and welcomed – as *heroic*. This is a good description of Occupy, which had many 'heroic confrontations with the authorities' and, arguably, sought to provoke them. The absence of leaders and icons is more than compensated by the bravery of the group. The confrontations with police, with the banks, with the elite, with the city, that they dare to take back and reclaim the city, or as Castells (2012) asserts, to reclaim democracy itself. Occupations themselves are a performance of confrontation, and constitute part of the narrative of heroic politics.

Change often arises as the product of confrontation or conflict. A central purpose of this book is to highlight the idea of confrontation and the many forms it may take, the diverse purposes it may serve. But with Occupy, there also is the idea that occupations in their public confrontation bring a simmering problem to a boil, and turn economic and social struggles into actual physical struggles with authorities. Among activists, even among those with the greatest optimism and strongest beliefs, there is the acknowledgement that the challenger, the upstart, the disrupter will lose that battle. Those who find such conflicts desirable or satisfying are nevertheless aware that such moments alone are not likely to change the direction of the struggle or achieve anything specific. These acts serve to inspire and to cohere the group (Graeber 2011). Half of what makes the actions heroic, in the sense of the nobility of the hero, is the willingness to engage in the face of such poor odds.

All these narratives contribute in different ways to a larger narrative of who the citizen is, in ways that are both progressive and problematic. Occupy mobilized further a story of 'a good citizen' among activists: someone whose bravery and willingness to confront demonstrates a real commitment to the movement, who has undergone a personal transformation of enlightenment, who recognizes the enormity of what is to be done and the importance of organizing to make it happen. To engage in heroic politics, these citizens allow themselves to feel outrage and allow it to energize them past the fear (see Castells 2012). These citizens are self-sacrificing and become part of something larger than themselves. But of course, not everyone is in a position to make that kind of contribution, nor to want to, and glorifying the more public frontline of a movement or event may lead to forgetting and devaluing the labour that supports the events and their participants. Graeber holds a fuller view of the complexity of activism's heroic actions, recognizing that the performance of activism that gets all the attention and comes to define politics is only the face of a larger process in which many more invisible people work behind the scenes to make the performance possible. He further notes that it is often women doing the hidden work, and the attention and legitimacy given to the performance once again makes their presence and contribution less visible and less valued (Graeber 2011). What is also clear from Graeber is that he is acknowledging and responding to the ongoing narration of heroic acts of confrontation and conflict. For some, that is a narrative of activism that Occupy preserves.

Occupy as art

When we talk about Occupy, we also refer to the symbols of Occupy, its iconography, the performance art of its protest, taking art as that which transcends the normal and everyday, expressively, and enables us to see differently and better understand the normal and everyday. But this is not only to speak of the art and poetry produced by activists as separate objects of study, as elements of a discourse; it is also to think through the way in which occupations are art forms themselves and what productive insight into politics may come from such a framing. Rancière's discussion of the relationship between art and politics, and the way each challenges the status quo, helps unlock some valuable distinctions in the representations of Occupy.

There is some evidence that occupiers and their supporters saw Occupy as art, in different ways. For Schneider, Occupy was 'a living work of art':

> a square block of granite and honey locust trees…became a *canvas* for the image of another world.…With artists in charge, Occupy Wall Street was art before it was anything properly organized, before it was even politics.
>
> (Schneider 2013: 5–6, emphasis added)

For Taussig, Occupy was a form of art in that it was the return of a type of politics that incorporated the sensory in a primary way: 'Politics as aesthetics is back. Politics as "affective intensity" is back too' (Taussig 2013: 40). He found a strong relationship between the words and images that commerce places all over the city, and the aesthetic of Occupy: 'For a century now, advertising signs and images have stolen from the avant-garde. Now it's payback' (28).

The role of art in its many forms within political movements is longstanding and complex. 'While they are not always self-consciously revolutionary, artistic circles have had a persistent tendency to overlap with revolutionary circles; presumably, precisely because these have been spaces where people can experiment with radically different, less alienated forms of life' (Graeber 2011: 98). For Rancière, both politics and art are forms of *dissensus*: 'a conflict between a sensory presentation and a way of making sense of it, or between several sensory regimes and/or "bodies"' (Rancière 2010: 139). 'Politics, before all else, is an intervention in the visible and the sayable' (37). Politics is specifically any act that disrupts the police order, which is often asserted and understood as

> the 'natural' order that destines specific individuals and groups to occupy positions of rule or of being ruled,...to specific ways of being, seeing and saying. This 'natural' logic, a distribution of the invisible and visible, or speech and noise, pins bodies to 'their' places and allocates the private and the public to distinct 'parts' – this is the order of the police....It [Politics] begins when they make the invisible visible, and make what was deemed to be the mere noise of suffering bodies heard as a discourse concerning the 'common' of the community.
>
> (Rancière 2010: 139)

Similarly, art creates something different, something previously unseen or unheard, 'a new partition of the perceptible' (Rancière 2010: 122). Despite this commonality with politics, it does not follow for Rancière that art is necessarily political, nor that 'art compels us to revolt when it shows us revolting things' (135). Through a critique of Deleuze's conclusion that 'art *is* politics' (172), Rancière argues that, although they are not one and the same, as they both imitate and contradict each other, neither can they be fully separated. Art's capacity for resistance lies in its assertion of art as art and only art, which is in itself also a refusal of politics, or of being political. This refusal is itself political, of course, but in becoming political, it is, strictly speaking, no longer art.

Mindful of these tensions, we can nevertheless salvage both an appreciation for the aesthetic qualities of protest in its performance and spectacle, whether explicit and intentional or inherent, and the necessarily mobile or shifting 'status' of an act moving between art and politics and back again. Framing Occupy as art allows us to see how it is an act of dissensus, despite its lack (in some instances) of leaders and demands. Even without a list of demands or goals, the phenomenon of Occupy still communicates dissent with some effectiveness.

The Occupy slogan, 'We are the 99%', is therefore perhaps best understood primarily as art, not politics. Chomsky wrote that this divide was among the 'things that were sort of known, but in the margins, hidden, which are now right up front' (Chomsky 2012: 70). What it achieves in the first instance is not a simple exposure of inequality to create rage to create political action, but a critical reframing of demographics and economics, which destabilizes the privacy and secrecy the '1%' previously enjoyed. As Rancière explained,

> Within any given framework, artists are those whose strategies aim to change the frames, speeds and scales according to which we perceive the visible, and combine it with a specific invisible element and a specific meaning. Such strategies are intended to make the invisible visible or to question the self-evidence of the visible…
>
> (Rancière 2010: 141)

Mitchell notes the lasting impressions of Occupy are somewhat unusual in what they have made visible:

> The iconic moments, then – the images that promise to become monuments of the global revolution of 2011 – are not those of *face* but of *space*; not figures, but the negative space or ground against which a figure appears. The only figure that circulates globally, that embraces both Tahrir Square and Zuccotti Park, is the figure of *occupation* itself. But occupation and the Occupy movement have no definite form or figure other than the dialectical poles of the mass and the individual: the assembled crowd and the lone, anonymous figure of resistance. And occupation, it should be noted, is not only a visual and physical presence in a space, but a discursive and rhetorical operation.
>
> (Mitchell 2013: 101–102)

The crowd in a public space may productively be seen as a work of art, as a performance that creates a 'we', and generates further creativity. The identity of that collective is complicated: the crowd is both newly visible (in its mass) and invisible (in its anonymity and indistinctness).

The art of Occupy, or Occupy as art, may be seen as mobilizing direct political action for some, but this is not the first dissensus work it does. As art, it brings to the surface and makes sayable the previously unspoken, unacknowledged reality of inequality (predominantly in terms of income). Moreover, there were no leaders – no generals or orators – nor great speechifying, and no one points to memories of such rhetorical persuasion as what Occupy was or means; this reflects Rancière's argument that art is also a refusal of 'the politics of orators and militants, who conceive of politics as a struggle of wills and interests' (Rancière 2010: 163). In this act of surfacing, Occupy is of course far from unique, although its celebrators sometimes exaggerate its effect, the reach of its impact and, most especially, its uniqueness. Their view is too exclusively political, and strategic, with too clear an agenda. It does not recognize the other work of Occupy. It misses seeing the crowd as an art form, a poem: a spoken word or dance or scream that exists for its moment of embodied performance – and then is gone.

Occupy as grammar

Here I would like to play a little bit with Occupy as grammar, and the idea of occupation discourse as acting like a grammar of protest politics. I take grammar as a system that identifies elements by the function they serve, and one that sets the relationships the parts must have to each other (order, change of case) in order (literally) to *make* sense. Can we see Occupy as a similar system for the expression but also the creation of meaning? It is through grammar that we understand that meaning is produced and comprehended through a stable, commonly understood system governing the relations between and among elements; meaning does not inhere in the elements themselves. A grammar constitutes the structure through which meaning will be produced and understood. In this sense, what does activism *mean*? What does Occupy tell us about the elements and meaning of protest?

I am not the first to use 'grammar' in this context. Nancy Fraser has made reference to 'a shift in the grammar of political claims-making' in post-socialist politics (Fraser 1997a: 2). Antonio Negri writes of seeking 'a new grammar of contemporary politics', but does not delve into the metaphor beyond the idea of a 'new vocabulary' (Negri 2008: 13). Cristina Bicchieri wrote of 'the grammar of society' in a more elaborate way that is closer to my intentions:

> Norms are the language a society speaks….I call social norms the *grammar of society* because, like a collection of linguistic rules that are implicit in a language and define it, social norms are implicit in the operations of a society and make it what it is. Like a grammar, a system of norms specifies what is acceptable and what is not in a social group. And analogically to a grammar, a system of norms is not the product of human design and planning.
>
> (Bicchieri 2006: ix)

Bicchieri's point is that it is not an explicitly deliberative process. She calls it instead a 'heuristic' process, and 'an automatic response to situational cues' (Bicchieri 2006: 5), although these are not completely opposite or separate from deliberation, in practice. These are codified rules that emerge from practice, usage, and occasional systematic thinking.

'Occupy' itself (by which I mean all that it represents and signifies) has become a set of structural rules that frames and structures protest and the position of protest within (and interaction with) the city, making visible the assumed rules and relationships of activist politics and the daily life of the city and its occupation. There exists a structure that silently supplies some of the logic for some of the above narratives. Because, as the previous aspects of discourse on Occupy have brought to the surface, there are existing expectations and 'rules' about political behaviour: who may be political, where, and in what way. These are based on identity, but in a way that is intimately connected to or rooted in ideas of politics and ideas of city space. Some of the evidence emerges in the selection of spaces by protesters for marches and for encampment, as the determination of their symbolic and logistical qualities is relative to those making the choices.

To occupy is also to busy oneself with something, and certainly the context of Occupy Wall Street or Occupy Dame Street was the concern of its participants not only for the immediate space of occupation, but with the other, usual activities of those places of financial and political institutional power. And as Harcourt also noted, occupation also has a military, strategic, territorial connotation; Angela Davis argued the purpose of Occupy was to push for the de-occupation of other sites by US military (Harcourt 2013: 70). Occupy is a transitive verb, so what is the direct object of 'Occupy'? What is *occupied*? The precise meaning of occupy is in part determined by its object. Central to its expression is the idea of occupation – this is the predicate. The unnamed subject is 'the people'. The verb's object in both an immediate and metaphorical sense clarifies the meaning of that verb. In most cases, the object of Occupy in its name was its host city, but the scale and lack of specificity of 'Prague' or 'Galway' or 'Mumbai' maintain a certain vagueness of the meaning of the occupation. Among all of this, perhaps we can see the occupation as trying to express something, and to make sense – which in the case of protest is commonly an effort to make sense in the face of something that does not make sense.

It may be useful to think of the functioning of ergative verbs here. Ergative verbs (generally rare and limited in English) are those whose action or impact on an object may be assumed by the agent as if there is no real agent, so that there is an implied transitivity, although the verb stands as intransitive: e.g. the bomb exploded. The subject is also the direct object. One effect of ergative verbs is to erase the cause of an action: things just happen. Although 'occupy' is not literally an ergative verb – 'Occupy Dame Street' takes an object, and lacks a subject only because 'occupy' is used in the imperative case as a command or an imploring – it has a similar sensibility of something happening rather than something being done to something or someone, because the subject or agent occupying is not explicitly specified. The common use of 'Occupy' as a noun, to refer to the name of the event or movement, perpetuates the idea of occupation as a thing that just happens.

This is not an uncommon sense of how mass mobilizations begin; as Graeber says, 'when a true revolutionary movement does arise, everyone, the organizers included, is taken by surprise' (Graeber 2013: 3). There are several references to the surprise of the 2010–11 uprisings. Castells opens *Networks of Outrage and Hope* with, 'No one expected it. In a world darkened by economic distress, political cynicism, cultural emptiness and personal hopelessness, it just happened' (Castells 2012: 1). Its agent-less spontaneity may not be an accurate description, given not only the explicit call by *Adbusters* that led to OWS, not to mention the general assemblies that occurred prior to it (Graeber 2013), but the appearance and the impression of 'it just happened' is a part of the discourse of Occupy as a form of politics.

Miriam Sitrin uses a biochemical metaphor of DNA to describe how she has observed the ways in which Occupy fundamentally changed the structure not only of protest but of the political subjectivity/sensibility of activists. After a list of material accomplishments, for Sitrin, the greatest achievement was something intangible: 'Most of all however, we have accomplished new relationships, reclaiming our relationships to one another and creating new ways of doing things, new ways of being....' Activism, for Sitrin, consists of the organizing, the networking, and the relationships, first and foremost, and the agenda and action emerge from there, rather than participants being drawn to a specific cause. In a grammatical frame, the structure enables (or is seen to enable) the expression, and it is ergative, in that Occupy, the verb-noun that is the object, becomes the agent that acts on its participants.

> Occupy is not a social or protest movement....No longer seen with the occupation of parks, plazas and squares, Occupy has relocated, it is in us, it is in our ways of being, relating and coming together....Occupy was never about a place or a moment – it was and is about a way of being and doing.
> (Sitrin 2014)

To look at this another way, just as we speak of transitive and intransitive verbs, is it productive to speak of transitive and intransitive politics or acts of citizenship? Our common understanding of politics is that it is inherently transitive – acts whose purpose is necessarily to cause another thing to happen, such as legislation. I would argue that Occupy is both transitive and intransitive citizenship in the way it has been understood and practised, and that this framework helps unpack the complexity of its politics. It concerns both being and becoming, of the individual and society.

The most interesting and problematic part of this grammar is the implied subject of a mass political occupation by 'the 99%': not just people, but 'the people'. In its specific combination of visibility (as a crowd) and invisibility (anonymity), the grammatical agent of an occupation is not a specific person, but 'the people'. Invoking 'the people' as Occupy (or any occupation) does, explicitly and implicitly, raises two questions, or signals two ideas: of democracy and of 'human rights'. These are related, but fundamentally different, understandings of politics: the discourse of 'human rights' as a basis for protest, and acts of 'citizenship'. In a human rights framework, rights do not arise from specific claims on or demands of a state; as a political framework, it asserts a way of being that pre-exists politics. To some degree, human rights purport to transcend politics. They are non-negotiable and nothing is supposed to override them. They pre-exist and thus trump other rights and they trump specific opinion. Such rights are declared to exist inherently and globally – they are generally the rights and freedoms deemed necessary for biological, social and political survival, as a defence against suffering. Occupy operates first as a human rights discourse and then as a citizenship discourse. It is asserted as a way of being, but in constituting the people, it also asserts citizenship rights in a democracy.

The idea of democracy and its attendant citizenship necessarily refers to a political exchange, embedded in systems of negotiation, debate, law and governance. Human rights discourse does not challenge the idea of 'the people', but rather reinforces the idea of a social body at increasingly larger scales. It is the political logic behind Hardt and Negri's assertion of/hope for the emergence of a global social body, or 'people'. However, the idea of 'the people' and its political system of 'democracy' are problematic for activist practices. Although they sound inclusive and reflective of the grassroots, they are myths put to work in the service of the powerful.

Occupations mobilize an art and grammar of the activist crowd; leaderless, demand-less Occupy only intensifies the importance of the crowd. Rancière explains the necessity of constituting collective identity:

> The 'aesthetics of politics' consists above all in the framing of a *we*, a subject, a collective demonstration whose emergence is the element that disrupts the distribution of social parts, an element that I call the part of those who have no part – not the wretched, but the anonymous.…It does not give a collective voice to the anonymous. Instead, it reframes the world of common experience as the world of a shared impersonal experience.
>
> (Rancière 2010: 141–142)

Discussion of the function of the idea of 'the people' is largely absent from English-language scholarship on democracy (Bonaiuti 2013; see also Rancière 1999) – interestingly, English does not even have a separate word for the concept like French (*le peuple/les gens*) or Italian (*il popolo/la gente*). In French- and Italian-language scholarship, the discussion is active and the argument is blunt: 'The people [*il popolo*]…is an invention' (Mastropaolo 2013: 23). As Bonaiuti has argued, 'The scandal of democracy consists of the fact that it is a political form of a subject who does not exist. "The people" do not exist…' (Bonaiuti 2013: 147–148). 'The people' only exist in the sense of a diverse and complex plurality of individuals, not as a singular collective. Much of a society's population, however, remains unrepresented and 'unreadable' in public life, unknown and invisible (Rosanvallon 2014: 17).

The language of 'the people' is strategic rather than descriptive, and serves to privilege 'hypostatizing non-dynamic forms of collective identity' that are non-inclusive and serve to hide the diversity and inequality among citizens (Bonaiuti 2013: 149; Rosanvallon 2014). Leaders speak of 'the people' – representation of them, love for them, commitment to them – in order to constitute discursively the polity that is governed, and to characterize it as unified. Its unity implies a coherence that circles back to justify the natural logic of the polity's existence. 'To put it another way: "the people" is an instance, not a substance…' (Bonaiuti 2013: 160).

Its strategic uses come from below as well as above. Those challenging the State in the name of the people represent the latter as a singular, indistinct body; but it is a false unity that ignores the antagonistic relationships among citizens:

> The people is an abstraction, which dissimulates a subject [*materia*, not *soggetto*] constructed first of pluralities – social, cultural, territorial/regional, generational, of gender, and so on. It is a togetherness of parts, which in the real world never move themselves in unison, but for the most part in opposition [to each other].
>
> (Mastropaolo 2013: 23–24)

The claim of constituting 'the people' is seen to legitimate the cause or the act of protesters, because the (vague) idea of democracy as government by the *demos* makes such collectivities authoritative or authentic (Mastropaolo 2013). Occupy's participatory deliberative practices add weight to the idea that Occupy is 'real' democracy, and thus more truly representative.

Resting Occupy on its '99%' is more unstable than it first appears. Colliot-Thélène (2011) challenges both the idea that 'the people' is a united, coherent, clear and stable entity, and that democracy has ever been based upon government by it. Like Bonaiuti, Colliot-Thélène takes a particular interest in democracy as it has actually and historically been practiced through citizenship, in the relationship between the governor and the governed. Representational government has left those who are governed far removed from the practice of governing; actual rule by the *demos* has never occurred, only a superficial democratization of the existing state structure (Colliot-Thélène 2011; see also Mura 2014). Moreover, representation needs to mean more than simply an assigned or elected representative who is accountable in some way; it is 'to see one's interests and one's problems publically taken into account, one's lived realities exposed' (Rosanvallon 2014: 35).

For all its problems, the idea of 'the people' is nevertheless resurging; its post-national or global applications make it more complicated, but have not slowed it down. Scuccimarra (citing David Tarrizzo) argues that increasingly 'it no longer makes sense, therefore, to ask if a certain political movement is or is not populist, because the political is populism' (Scuccimarra 2013: 12). 'The people' is precisely the language Castells uses to describe this process in the 2010–11 occupations: 'individuals did come together again to find new forms of being us, the people' (Castells 2012: 1). As noted earlier, Solnit identified the same process, when she writes, 'In these moments of rupture, people find themselves members of a "we" that did not exist, at least not as an entity with agency and identity and potency, until that moment' (Solnit 2013: x). When their act of 'framing of a *we*' becomes political within a democracy, the crowd claims to constitute and *represent* the nation. When Chomsky writes, that Occupy 'should be regarded as a response, *the first major public response*, in fact, to about thirty years of a really quite bitter class war that has led to social, economic and political arrangements in which the system of democracy has been shredded' (Chomsky 2012: 54, emphasis added), in his use of 'public' he is arguing that Occupy has constituted (or at least started to constitute) 'the people'. The other, less-visible (to him) responses do not represent 'the public' or 'the people', as he argues that

Occupy does. However, in no way – not democratically, nor demographically, nor socially – does Occupy or any other mobilization represent 'the people'.

The occupied city

As the direct object of the occupation, when we talk about Occupy, we also talk about the city and its proper uses. In the ergative grammar of occupation, the city also occupies its participants and occupies itself. Occupying public space turns a space designed for usage rights into a territorial claim, however temporary. The challenge makes visible the territorial claims that exist, and brings to the surface who actually owns or governs the spaces we see as common and collective. Occupations may also make visible those who are otherwise invisible, occupying not only material public space, but the public eye, thus representing the city (however limited, however briefly) to itself and the outside world. The city lends itself to occupation, to being seen, with the presence of newsmedia as well as individuals' and the State's omnipresent cameras, not to mention the way the mirrored glass of its buildings reflects us back to ourselves. The city increases, enhances and complicates the degree to which we are visible to each other and to ourselves.

While occupying disrupts public space and draws attention to contested claims on it, inhabiting speaks to both public and private use of the city, the former for temporary usage, the latter for more permanent claims. Inhabiting, as an embodied and lived experience, is a much more temporary and ephemeral practice than we typically depict or understand it to be. While inhabitation is often grounded in and centred around residence, that residence itself may be mobile or regularly relocated. Moreover, inhabiting a city is not only physical residence, it is also necessarily a series of movements through and interactions with the city's physical and social landscapes, without permanent claims or traces.

When we talk about Occupy, we talk about much more than a specific city or movement or march or occupation. The discourse of Occupy is one of the key vocabularies of politics. It is not the only one, but its discourse is highly indicative of political diversity and divides, within the academy and within the general public. Its discourse reveals our understandings of the purposes of cities and their specific places and the role of politics in our daily lives. While there are clearly gestures towards the financial and political power of cities, does Occupy otherwise present the physical space of its host cities as a neutral space, a stage for the performance of protest? The places that are occupied are not residential; they are generally spaces people pass through. Does occupying such places in effect perhaps make less visible (however unintentionally) the crisis of the city itself, particularly as it occurs *prior* to the global financial crisis? Even though encampments regularly included the deeply marginalized who do inhabit parks and public places, does the occupation make the inhabiting less visible?

It is valuable to take inspiration from Harcourt and Davis in thinking about other meanings of 'occupy': many would argue the city was already occupied, and the city already in crisis. Not to mention the climate crisis, several refugee crises, or the crisis of democracy some have identified in the wake of the EU's interventions in the Greek and Italian governments. Which – or *whose* – crisis is highlighted by Occupy movements? Castells recognizes the geographical limits of the global financial crisis and demonstrates how it has come to dominate certain political agendas, of both the State and activists:

> However, beyond the shores of North America and Europe, most of the world is *not* in crisis – at least, not in the crisis of global capitalism that shook up the until-now dominant economies. Granted, poverty, exploitation, environmental degradation, epidemics, widespread violence, and uncertain democracy are the daily lot for people living in what are known as emerging economies.
>
> (Castells et al. 2012: 12–13)

The 'daily lot', even for many people in *advanced* economies, is in crisis: many indigenous peoples, African Americans, Travellers, Roma, refugees in 'The Jungle' in Calais, people with insecure housing, to name only a few examples, are facing life and death struggles as part of their everyday life. Occupy invokes a politics that does not capture fully the diversity of the inhabited city, their crises, nor the political response from those facing those crises.

I argue for using the inhabited city, rather than the occupied city, as our starting place for radical politics – although they are not actually separate: organizers make use of the city for protest, not just in occupying a park, but meeting up in cafes, restaurants, and in open public spaces, for planning protest as well. Occupations are spectacles that appear to be and are sometimes celebrated as something other than they are. How occupation is seen in the first instance as an act of politics is often a reflection of the authors of those opinions who see the next rebirth in the cycle, where 'we' will once again be authentic, finally start the revolution, constitute 'the people', and heroically lead the charge. But a more thorough look reveals other work occupations do: their contributions, which act in a more ephemeral way, and their limitations – the narrowness of the frame and the impossibility of representation, their neglect or exclusion of the inhabited city and its daily injustices. Occupy is significant, but its greatest value may be in its perceived weakness – its failure to lead, to last. We must look beyond it, too, to find a politics that includes and is expressive of that diversity of the invisible inhabited city.

We need to see more of the city and its activism than certain readings of Occupy show us. Embedded in the comments of those who see a new politics and potentially revolutionary change in Occupy are ideas about the city, and ideas about whose politics matter. To believe that revolutionary change looks like Occupy is a reflection of a particular kind of citizenship rooted in a particular kind of inhabitation of the city. Seeing Occupy, even as a global phenomenon, as *the* awakening, or the start of a general strike, the start of some kind of revolution, the first sustained response to anything, echoes the 'state-sanctioned, normalized White, middle- and upper-class, male heterosexuality' theories of politics that scholars like Cohen are trying to move beyond (Cohen 2004). One does not have to actually be all of white, cis-male, straight, enabled/able-bodied, settled, middle-class and to take that position, but it is only from that perspective of analysing and, more significantly, *inhabiting* a city that one could produce such a politics. A politics that fights the oppression, exploitation, and injustice of global neoliberal capitalism in isolation is a politics of the citizen who is a white, cis-male, straight, enabled/able-bodied, settled, middle-class activist. That citizen inhabits the city in an otherwise privileged fashion. Nothing else thwarts or threatens his existence, his daily life, his right to speak.

I struggle with the celebratory, self-congratulatory, triumphant depictions of Occupy. Occupy was important as dissensus, for what it made sayable and for how the experiences were inspiring and transformative for its participants and observers. It was a festival moment of gathering, joy-making, creating, and pushing the boundaries. It is also important to recognize that it was one event in a chain of other events, already begun, with no discernible beginning. It was a more polished performance of the politics that underwrites day-to-day struggles. Yet in other ways, it was quite unpolished, imperfect, even tarnished. To describe it as art, to see its affinities with dancing, flash mobs and sculpture, is not to reduce or diminish it, but to reframe it and reveal the work it did and continues to do, in both progressive and problematic ways.

In the pages that follow, I make a case for framing activism in terms of citizenship that is grounded in the inhabited city. Citizens, as individuals and in groups, are what constitute the 'people' in a differentiated manner, rather than as a mass or multitude. The ideal of citizenship is a political identity/subjectivity that alleges to transcend all others and establish equality in the citizen's relationship with the State. In its historical practice, citizenship has always been employed as a differentiating practice, a means to exclude and privilege. No Western nation-state has ever lived up to its stated constitutional principles. Those constitutional documents, however, leave a space open for leverage against those unmet promises and new forms of constitutionalism that are enacted, rather than represented.

We need to talk about more than Occupy. Just as importantly, when we do talk about Occupy, we need to become more aware of the diversity of understandings it has mobilized, as well as recognizing that it has been used to privilege certain ideas, assumptions and expectations. There is no such thing as 'the people'. Mass mobilizations when comprised of 'citizens' frame activism in a way that does not rely on any given democratic structure, nor overstate the capacity of the crowd to represent other citizens. This is not to dismiss highly visible activism, but also not to privilege its participants and allies. Occupy was an act of dissensus and an act of diva citizenship, among many others. As we think about our own expectations of what Occupy or any other protest is supposed to accomplish, and the larger political context of a crisis on the Left in which activism is often situated, the limitations of taking Occupy as a model become clearer still.

2 Radical politics and the 'post-political' critique

Toward the end of Sean O'Casey's 1924 play, *Juno and the Paycock*, Maisie Madigan, the older, meddling, upstairs neighbour, shouts at the police as they depart her tenement:

> For you're the same as yous were undher the British Government – never where yous are wanted! As far as I can see, the Polis as Polis, in this city, is Null an' Void!
>
> (O'Casey 1924: Act 3)[1]

Set in 1922, the play documents life for a family in Dublin at the start of the brief Irish Civil War following the War of Independence. Similarly to O'Casey's *The Plough and the Stars* (set during the 1916 Easter Rising), the play raises questions about the cost of revolution and war for ordinary people, and challenges simple narratives of positive change. In *Juno and the Paycock*, families are struggling economically, and dealing with losses and injuries from the conflict, with no obvious benefit or improvement from the defeat of the British. Much of the characters' subsequent suffering comes not from bosses or governors, but rather at the hands of family and friends in the form of betrayal, abandonment and rejection, and is in large part beyond any state to prevent. Mrs Madigan's frustrated outburst at the police articulates an argument that the change of government from British to Irish has not increased the security of ordinary citizens (with the fantastic added touch of the double meaning of *polis* as accented 'police' and as city). One crushing blow to the family, the loss of an anticipated inheritance, turns on a minor legal error of wording, suggesting that for the marginalized, even the law is a fragile, imperfect and thus unreliable measure of protection and security.

O'Casey's purpose, and mine, is not to erase all significance of the overthrow of British rule in Ireland, or any other political revolution. It is to highlight the significant limitations to even revolutionary change of formal political institutions. Key institutions such as the police and the law may not and frequently do not serve ordinary people well, regardless of the form of government; they often serve to entrench power hierarchies and the authority of the police. As Scott noted, 'The revolution…is rarely if ever the end of peasant resistance' (Scott 1985: 302). Moreover, oppressive systems such as patriarchy and white supremacy may continue to afflict suffering under any regime, democratically elected or otherwise.

What are our revolutionary goals? How do we measure or assess change that is a success? What difference does 'success' make? For whom? The discourse of Occupy shows that this is an ongoing debate about goals, strategies and tactics for revolutionary activists. While we are well aware that a change in government, even a revolution, does not change everything for everyone, much political theory continues to privilege these formal, institutional transformations as having greater impact and as the model for measuring major change, and it privileges formal political change as a 'bigger' or more relevant form of change. Historians (not only political ones) frame temporal categories around them, such as the 'Revolutionary Era' of the late eighteenth century (which itself privileges political revolutions in the US and France over Russia's).

Determining what kind of change is needed (and how to achieve it) is a central preoccupation of the Left. Many radical political theorists see themselves as 'writing in the wake of the disaster of state socialism. They are searching for a way to reinvigorate the project of developing critical alternatives to capitalism, even as they categorically reject the state-socialist alternative…' (Purcell 2013: 35). Activists and scholars on the Left largely agree that, for a variety of reasons, the Left has been outmaneuvered: its issues have become beyond debate, or 'post-political'. Insight into mechanisms of the foreclosure of political debate and subsequent proposals for new frames of politics as a way forward have come from literature critiquing the emergence of a 'post-political' era, an era that has frustrated efforts on the Left to define and assert itself, especially in the context of the post-Cold War era and neoliberalism. This work nevertheless has yet to engage fully with intersectionality and the limits of its own perspective, and thus proposes change that would leave unaffected many forms of suffering, injustice and oppression. As Purcell has observed, 'unfortunately class reductionism remains, even today, a bad habit of thought among many on the left' (Purcell 2013: 62). Critics of the 'post-political' also follow chronologies specific to a particular view, anchoring their analyses around moments assumed to have universal significance. Within those chronologies, certain crises are highlighted while others are misrecognized as 'new' or missed entirely.

In the seemingly impossible context of the 'post-political', debates continue about what is to be done, what would constitute success and failure. This chapter is not a discussion that seeks to determine a measurement for success or failure, or one that presumes to judge any given campaign, movement or event. Rather, it is an inquiry into the nature of that measurement and judgement, the ideas of politics and citizenship in which such ideas are rooted, and the ways in which such frameworks continue to shape our expectations of ourselves and others, and of the possibilities for change. Reviewing the work of Chantal Mouffe, Nancy Fraser, David Graeber, Franco Berardi, and

Mark Purcell in this debate, I discuss the strengths and weaknesses of efforts to 'restart' the Left in a way that addresses diversity, in light of frameworks from such scholars as Kimberlé Crenshaw, Patricia Hill Collins and Iris Marion Young for thinking through the difference that difference makes.

The unbearable whiteness of the post-political critique

The post-political literature has sought to understand how it is that more and more political ideas and acts are deemed 'impossible' and undebatable. This critique is astute in its depiction and analysis of neoliberal, especially post-1989, political strategies by governing parties. It contextualizes 'New Labour' and the movement of social democratic parties away from more 'radical' agendas towards strategies for electoral success (and how their traditional social justice campaigns came to be termed 'radical', even among their own membership). Neoliberal policy efforts to transfer issues from the public sphere and into the realm of private and individual responsibility have made it difficult to raise such issues in public debate as they are no longer considered to be subject to it. This has been accompanied by a move towards technocratic and managerial solutions, rather than explicitly political or state-led action, fed by the logic of the 'end of history' (Fukuyama 1992) and Thatcher's language of 'there is no alternative' to the market governing economic relationships and private responsibility governing social relationships (see Mouffe 2005 [1993]). Fukuyama's theories and the death of society have been widely debunked in many of their details, but there lingers a powerful sense that liberal democracy and market capitalism in some form are the inevitable means through which we will deliberate and administer the distribution of resources. Neoliberalism in practice has alternated between a strong presence and an active absence on the part of the State, from deregulation and tax cuts, to bank and other corporate bailouts. The contribution of the media to this sensibility has repeatedly been called into question (Herman and Chomsky 1988; Mercille 2014, 2015) and Graeber also cites the related 'manufacture of intellectual authority' as one cause as to why 'real political debate becomes increasingly difficult' (Graeber 2013: 117). Rancière argues that by rendering dissenting voices inaudible by deeming them illegitimate, elites have denied those voices recognition as political subjects (Rancière 1999, 2010). At the same time, intensified and increasingly militarized police action has become a response to, but also its own evidence of, the danger of challenges to the status quo (L. Wood 2014). The cumulative effect is to physically and discursively shrink the space in which counterclaims can be made, to delegitimize challenges to the status quo including nearly all forms of protest – particularly anything deemed 'uncivil', much less 'violent'.

At the same time that this neoliberal, post-socialist moment was identified, scholarship in several disciplines was articulating a strong case for the incorporation of identity-based discrimination into all forms of social analysis. A landmark article on intersectionality by the legal scholar Kimberlé Crenshaw drew on Black feminist theory (e.g. hooks 1981; Hull et al. 1982) and explicitly challenged the 'single-axis framework that is dominant in antidiscrimination law and that is also reflected in feminist theory and antiracism politics' (Crenshaw 1989: 139). Through the grounded and specific frame of law and the courts, Crenshaw exposes the hypocrisy that underlies preventing Black women from representing all women (something white women are allowed to do), while denying Black women recognition of their specific experience. She argues that within any given group, the most privileged are taken as normative and representative, thereby erasing the more complex experience of those who are 'multiply-burdened'. Crenshaw emphasizes that, 'These problems of exclusion cannot be solved simply by including Black women within an already established analytical structure' (140). As axes of identity intersect, the result is not an arithmetic sum of identities, but a more complex social matrix of discrimination. This situation encourages individuals to 'guard their advantages while jockeying against others to gain more' (145).

This latter point is critical for recognizing the way in which purported allies may ignore or even sacrifice their less-privileged colleagues, and reproduce dynamics of oppression. In a subsequent paper on intersectionality and violence against women, Crenshaw elaborated on how the experiences of Black women activists may be conflicted and exhausted by the 'need to split one's political energies between two sometimes opposing groups…' (Crenshaw 1991: 1252). Moreover, the failure to acknowledge the Black woman's specific experiences of sexism and racism, combined with the normalizing of the dominant experience, often means that Black women's issues rarely set the agenda in either women's or anti-racist groups, and instead find themselves subordinated by a group's strategy for empowerment. More critically, intersectionality reveals the complexity of race and gender identities and the materially different treatment of both Black and white assailants and Black and white victims of violence: while all women are subjected to further trauma by a legal system that does not protect or believe them, Black men are assumed to be more violent and their violence more offensive when directed at white women, and Black women's trauma is taken less seriously and given less attention and support. In advancing her argument, Crenshaw also clarified that 'intersectionality is not being offered as some new, totalizing theory of identity' (1244). Its purpose is to insist on the constant 'need to account for multiple grounds of identity when considering how the social world is constructed' (1245). The politics of the Left, among activists and academics, does not transcend or stand outside its own social world.

Building directly on Crenshaw, sociologist Patricia Hill Collins (1990, 1998) argued that the particulars of Black women's intersectional subjectivities produced specific theories of justice and practices of resistance – which deserved to be recognized as such and not dismissed as 'thinking' in contrast to white men's 'theory' (see also Christian 1987). She points to the historical record and argues, 'the legacy of struggle among US Black women suggests that a collectively shared Black women's oppositional knowledge has long existed.

This collective wisdom in turn has spurred US Black women to generate a more specialized knowledge, namely, Black feminist thought as critical social theory' (Collins 1990: 12). She further argued that this applied to all groups – that social and political subjectivity were inter-related, embodied processes. Both Crenshaw and Collins remind us that white men's theory, although rarely acknowledged as such, also arises from embodied experience and is specific to their race, gender, sexuality, and so on. These differences are salient in the application of (and, thus, meaning of) justice: 'Under distributive paradigms of justice in which everyone is entitled to the same bundle of rights, for oppressed groups, diluting differences to the point of meaninglessness comes with real political danger' (Collins 1998: 149).

In political science, Iris Marion Young drew on feminist, postcolonial and critical race theory to argue for new theories of justice that went beyond liberal ideas of equality and beyond redistributive models. Identifying exploitation, marginalization, powerlessness, cultural domination, and violence as the 'five faces of oppression' (which may overlap), she argued for a less individualized approach to equality and justice that would recognize the experience and oppression of social groups *as groups* (Young 1990). She also examined how deliberative processes and means of communication can serve to privilege and exclude, and create a consequent need for protest for the excluded (Young 1997b, 2000). Although Young did not concern herself specifically with the post-political question, she directly engaged the debate regarding a way forward for the Left, particularly with Nancy Fraser.

Chantal Mouffe and Fraser were explicitly interested in challenging the 'post-political' muting of the Left, and both address the question of how to include difference beyond class difference into our political theory, as well as what it means for democracy, or how we should define/structure democracy so that it takes difference into account. In 1985 Ernesto Laclau and Mouffe had set out a theory that politics was always constituted by competing efforts at hegemony and thus, 'the observed political pluralism in society cannot be resolved into a homogenized social whole' (Laclau and Mouffe 2001 [1985], cited in Purcell 2013: 59). They argued for building solidarities they term 'chains of equivalence' with different groups coming together through the identification of common enemies.

Following the breakup of the Soviet Union, Mouffe specifically argued that as 'totalitarianism' had been 'defeated', democracy as a movement needed a new enemy (for the right, she says it is immigration). However, this enemy was to be accepted and engaged, not 'destroyed' (Mouffe 2005: 4). Consequently, she insists on the need to move away from the desire for consensus ('fatal for democracy'), and instead to recognize, expect and explicitly accommodate difference and pluralism, and the antagonism this necessarily produces – indeed, that is the very definition of politics. When Mouffe speaks of democracy in practice, she means liberal, representational institutions, even as she recognizes that the political exceeds those institutions and is embedded in every relationship and scale of human society, formal and informal. The irreconcilable diversity of polities must be acknowledged and embraced:

> [W]e have to break with rationalism, individualism and universalism. Only on that condition will it be possible to apprehend the multiplicity of forms of subordination that exist in social relations and to provide a framework for the articulation of the different democratic struggles – around gender, race, class, sexuality, environment and others....It means acknowledging the existence of the political in its complexity...
>
> (Mouffe 2005: 7)

Efforts to ignore or paper over this diversity and complexity, in search of 'neutrality grounded on rationality', are futile and tend towards fascism, because the idea of neutrality immediately constitutes a situation void of the political, and imagines the maintenance of order as apolitical (Mouffe 2005: 139). The more order is prized – the *liberal* over the *democratic* – the more it leans towards fascism. Mouffe's proposal is to embrace the tension between the liberal and democratic and make it productive (on this inherent tension or 'paradox' within liberal democracy, see also Mouffe 2009, 2013).

Fraser responds to the same pressures and argues that, post-1989, scholars and activists found themselves in 'the "postsocialist" condition':

> progressive struggles are no longer anchored in any credible vision of an alternative to the present order. Political critique, accordingly, is under pressure to curtail its ambitions and remain 'oppositional.' In a sense, then, we are flying blind.
>
> (Fraser 1997a: 2)

Fraser was unnerved by what she saw as 'the tendency to focus one-sidedly on cultural politics to the neglect of political economy' (Fraser 1997a: 174). She feared that the two camps, as she saw them, were missing the opportunity to draw on the strengths of each:

> Within the discipline of political philosophy, for example, theorists of distributive justice tend simply to ignore identity politics, apparently assuming that it represents false consciousness. And theorists of recognition tend likewise to ignore distribution, as if the problematic of cultural difference had nothing to do with that of social equality.
>
> (Fraser 1997a: 189)

Fraser sought to join both theories of justice because 'justice today requires *both* redistribution *and* recognition' (Fraser 1997a: 12, emphasis original). She acknowledged the violent histories that give rise to social justice movements (although she does not centre this violence), as well as 'the masculinism, the white-Anglo ethnocentrism, the heterosexism – lurking behind what parades as universal' (Fraser 1997a: 5). Mouffe also reminds us, 'We are all taught that the "individual" is a universal category that applies to anyone or everyone, but this is not the case. "The individual" is a man' (Mouffe 2005: 13). Fraser's working definition of identity speaks to intersectionality: 'social identities are exceedingly complex. They are knit from a plurality of different descriptions arising from a plurality of different signifying practices. Thus, no one is simply a woman…' And neither are they 'constructed once and for all and definitively fixed', but rather 'discursively constructed in historically specific social contexts…' (Fraser 1997a: 152). Similarly, Mouffe recognizes that, in practice, one of the problems with universalist democracy is that it simply is not what it claims. 'Social agents' are not 'homogeneous and unified entities' (Mouffe 2005: 12).

Neither Mouffe nor Fraser address how the privilege of socially dominant groups has accumulated over generations in material and discursive ways. The 'individual' (or, equally, the 'citizen') is a man, and is also white, a member of a dominant ethnic group, financially secure, heterosexual, permanently housed, cis-male, and able-bodied. Each of these identities and memberships has facilitated political participation in democratic processes, as well as economic access: jobs, promotions, safe and affordable rental accommodation, mortgages. Fraser is openly disappointed with the way 'identity politics' was used to dismiss and demean, and asserted that a new critical theory 'would need to take as its starting point the multivalent, contested categories of privacy and publicity, with their gendered and racialized subtexts.…It would also need to show how some of these publics marginalize others' (Fraser 1997a: 118). However, she does not address how such a starting point might change her overall project.

Although Fraser's goal is to integrate redistribution and recognition, her analytical approach is to address them separately:

> In the real world, of course, cultural and political economy are always imbricated with each other, and virtually every struggle against injustice, when properly understood, implies demands for both redistribution and recognition.…Only by abstracting from the complexities of the real world can we devise a conceptual schema that can illuminate it.…[R]edistribution and recognition [are] two analytically distinct paradigms of justice…
>
> (Fraser 1997a: 12–13)

Fraser's analytical framework thus works against her stated approach to identity. At worst, her divide creates no analytical space for intersectionality; at best, it considers intersectional identities to be the most problematic, rather than acknowledging that every identity is intersectional.

Young was critical of Fraser's dissection and reduction of injustice into only two categories, and disputed the claim that cultural theorists were ignoring political economy. Young countered that asserting political economy as distinct from and in opposition to culture misrepresented identity politics and social movements, and only made coalition-building more difficult. The 'contradictions' Fraser identified between redistributive and recognition struggles were, for Young, unfair distortions that could potentially mute already marginalized communities and even 'fuel a right-wing agenda' (Young 1997a: 160). She reiterated the goal of Cultural Studies, 'to demonstrate that political economy, as Marxists think of it, is through and through cultural without ceasing to be material....Political economy is cultural, and culture is economic' (Young 1997a: 154; see also Lovell 2007 on Fraser and materialism). She concludes, 'resistance has many sites and is often specific to a group without naming or affirming a group essence' (160).

'Iris Young and I seem to inhabit different worlds', responded Fraser (1997b: 126). She insisted that her categories of redistribution and recognition were not intended to map onto groups or social movements, but to abstract different 'paradigms of justice' that she saw as working at cross-purposes (Fraser 1997b: 127). Judith Butler added her thoughts to this ongoing discussion in *New Left Review* by tackling the weak understanding of 'cultural' that was circulating, i.e. 'that new social movements are "merely cultural"', allegedly lacking any sense of materialism. She raised the question of 'which movements, and for what reasons, get relegated to the sphere of the merely cultural...?' and was concerned that 'cultural' and 'material' are discursive tools employed 'to make questions of race and sexuality secondary' (Butler 1998: 36). Moreover, she disputed the framing of recognition politics: '*there is no reason to assume that such social movements are reducible to their identitarian formations*' (Butler 1998: 37, emphasis original). Fraser's response to Butler made several clarifications, such as that misrecognition was material and not 'merely cultural' (Fraser 1997b). These exchanges deepened and enriched the discussion of identity and put such issues 'on the agenda once again' (Fraser 1997b: 149), but Fraser did not take up the larger question of intersectionality, nor, more significantly, how it might shift the political frame.

Echoing some parts of the above exchanges, I want to discuss some weaknesses in the post-Cold War theorizing of the Left by Mouffe, Fraser and others, even as they tried to incorporate 'identity politics': first, their chronologies and, relatedly, their description of the 'post-political' foreclosure of certain politics as 'new'; and second, their ongoing reluctance to understand and engage 'identity politics' in any depth. Despite Fraser's fears, 'identity politics' did not swamp and wash out work on redistribution (see also Springer 2014; Gibson 2014).

The chronological frame of the critique of the 'post-political' is the rise of neoliberalism beginning in the 1970s, the end of the Cold War and the global financial crisis. However, these moments do not hold the same significance for everyone, in terms of identifying a source of a crisis or defining a political agenda. Many activists were and are challenging neoliberal policies, but also policies and crises that pre-date the 1970s significantly. Just as the 'global financial crisis' was not actually 'global,' affecting all countries (Castells 2012), the economic and political development of many communities does not map on to a Cold War chronology. While 1989 may have some resonance and post-2009 austerity programmes intensified suffering in many quarters, these have had not more impact, and in some cases less, than other watershed moments that shifted the political frame. For example, for many African Americans, the beating of Rodney King in 1991 and the acquittal of the police officers responsible the following year was a shattering of any trust in the system.[2] That such police brutality could remain unpunished in the face of video evidence exposed to a worldwide audience confirmed yet again that violent racism had survived the Civil Rights Era. 'Rodney King moment' as a phrase became a signifier for such moments of becoming 'woke', as they happen for each successive generation. As Cohen wrote, the abandonment of African American residents of New Orleans following Hurricane Katrina in 2005 was another:

> For this generation of young black people, Katrina is their 'Rodney King moment,' that visible rendering of black people and the black body as expendable, especially in the eyes and behavior of the state.…not only the physical beating of yet another black man at the hands of the police…but also the vindication of those white police officers.…Both components of these events served as a reminder to black Americans across the life course that although the formal laws of the United States may have changed, the ideologies and instruments of brutality used historically against black people were still employed and available to those in power.
> (Cohen 2010: 110–111)

The chronology of American racism, particularly issues of police brutality, is a straight and unbroken line back to slavery. The present challenge is the most recent iteration of resistance going back several hundred years. Beginning in 2013, the killings of Trayvon Martin in Florida, Eric Garner in New York, and Michael Brown in Ferguson, Missouri – all unarmed – generated another moment of intense grief and injustice that reset the political frame and agenda. Three Black women who work as community organizers launched Black Lives Matter, first as a hashtag on social media in response to the acquittal of Martin's killer, and then as a broader social movement following mobilization in support of the Ferguson community. There are now Black Lives Matter chapters in several other countries. Many neoliberal policies regarding drugs and incarceration, de-industrialization and gentrification, and de-investment from social housing and public education also frame the urban crisis for marginalized communities, and the timeline of their implementation is not irrelevant. Nevertheless, these policies provide an incomplete chronology for understanding the turning points in political thinking and activism for Black people in the United States and elsewhere.

For Purcell, the present moment has multiple concurrent framings: it is the product of globalization, neoliberalism after Keynesianism, the failure of an alternative from the Left, urbanization, post-Cold War and now war-on-terror security politics, and the climate crisis. Purcell zeroes in on urbanization as a global phenomenon, which is 'a process driven almost entirely by urbanization in the global South', which in turn is the result, in part, of deindustrialization of the North (Purcell 2013: 15). Rapid migration to cities often produces informal settlements, which have their bad and good sides: 'Informal settlements are of course places of great hardship, poverty, disease, crimes, and unemployment, but they are also, in a way, places off the grid, places where people are able to experiment with possible alternatives' and 'create something different, something other than Fukuyama's end of history' (17). This is a richer, and more complex chronology at multiple scales, and Purcell is drawing our attention to the urban in particular, whose suburbanizing, de-industrializing crises began (at least) in the early post-Second World War era. Still, this approach is from a bird's eye, 'universal' perspective, which needs to be complemented by the details of the inhabited city and the ways it frames political thinking and action.

The chronologies of the post-political-critique literature focus on the recent past, and this is directly related to another problematic aspect, which is the idea that identity-based social movements are new. Mouffe, for example, asserts that 'The new rights today are the expression of differences whose importance is *only now* being asserted, and they are *no longer* rights that can be universalized' (emphasis added, Mouffe 2005: 13). She argues that this is in part due to the vacuum of the post-socialist moment: 'Where there is a lack of democratic political struggles with which to identify, their place is taken by other forms of identification, of ethnic, nationalist or religious nature' (6). This approach fails to see the historical depth of those struggles (particularly nationalist separatism) and believes too much in the capacity of civic identities to supplant or transcend other identities for groups. There are instances in which groups are not actually capable of participating in larger 'solidaristic' political movements, not only because it is often not in their interest to do so, but because they have been actively excluded from the national/citizenship/civic identity that underpinned the political frame in which there was just 'left' or 'right'.

Even Graeber writes of 'The growing sense, on the part of Americans, that the institutional structures that surround them are not really there to help them…', as if alienation from political and social institutions were new. He identifies it as a class phenomenon, but noting that in the United States, 'middle class' is a cultural category, even a sensibility, more than it designates an economic group: 'It has always had everything to do with that feeling of stability and security that comes from being able to simply assume that… everyday institutions like the police, education system, health clinics, and even credit providers are basically on your side' (Graeber 2013: xxi). This is an important insight into class in the US, and resonates with Cohen's observation that there are those who are able to choose their interactions with the State, and those who cannot (Cohen 2004). But it also reveals the limitations of class-focused language on citizenship and interactions with the State.

Blacks, Latinos, gays, lesbians, disabled persons, the trans community, Roma, Travellers – none of these would say that any of those 'everyday institutions' were 'basically on your side', even if one's profession and income qualified as middle-class. The ways in which intersectionality complicates class are critical to understanding the actual practice of citizenship – both in terms of how individuals and communities are seen, and how they actually live in cities.

Franco Berardi's description of the current political moment also favours chronologies that create a historical break that only exists for some. He is less concerned with a post-Cold-War framework, focusing instead on neoliberalism, media control and the technology of the Internet: 'Society has been broken up, rendered fragile and fragmented by thirty years of perpetual precarization, uncontrolled and rampant competition, and psychic poisoning produced and controlled by the likes of Rupert Murdoch, Silvio Berlusconi, and their criminal media empires' (Berardi 2012: 49). Although neoliberalism worsened things the precarity (in multiple senses) of many groups pre-exists neoliberalism. For many, fragility and vulnerability have been a longstanding status quo.

Berardi also implies a new fragmentation of society, not just within the Left. The solidarity he envisions is 'based on the territorial proximity of social bodies', but it 'is difficult to build now that labor has been turned into a sprawl of recombinant time-cells, and now that the process of subjectivation has consequently become fragmentary, disempathetic, and frail' (Berardi 2012: 55, 54). Nostalgically, he remembers, 'Industrial workers experienced solidarity because they met each other every day and were members of the same living community who shared the same interests, while the Internet worker is alone and unable to create solidarity…' (118). While there are newly splintering forces at work, Berardi glosses over the long history of the many ways in which we do not 'share daily life' that have little to do with labor and everything to do with relationships of power rooted in distinctions of settler/indigenous, 'race', ethnicity (including nomadic peoples), gender (including pan and transgender identities), sexuality, and dis/ability. Berardi goes so far as to mark the present moment as a nadir for solidarity: 'Capitalism has never been so close to its final collapse, but social solidarity has never been so far from our daily experience' (59). Berardi's relunctance to nuance the impact of the Internet and recognize the inclusion it has generated for some workers (such as disabled people, whose Internet connection creates access to jobs and communities they cannot otherwise reach easily or at all), is only superseded by his erasure of the women, Blacks, Asians, Italians, gays and Travellers, and so on, who were not welcome in the workplace nor its union by their sexist/racist/nativist co-workers (see, e.g., Georgakas and Surkin 2012).

The idea that post-socialist identity politics, fragmentation of solidarity and the post-political foreclosure of dissent are new phenomena is problematic on several fronts. The only groups for whom these are new are those who were in the circle of formal, institutional politics and suddenly found themselves with less ground under their feet. The idea that one's proposed political, social and economic agenda is 'radical' and 'impossible' is the longstanding norm for many groups. Framing the Left's timeline as newly challenged by the post-political participates in the papering over of a rich history of resistance and rebellion by groups who also consider themselves part of the Left (We Are the Left 2016), and who have specifically integrated identity politics with Marxist strategies of redistribution (Georgakas and Surkin 2012). It presumes that differences (and violent hierarchies) were neither previously recognized nor asserted. Women who struggled for the very liberal-democratic vote were condemned as radicals and traitors to their sex, their families and their country, and were arrested and assaulted by police forces. In other words, they met exactly the same fate as many protesters today. Blacks arguing for the end of slavery, for the end of unequal segregated spaces, for voting rights, and for the end of police brutality have been greeted in exactly same way, both in their public and private treatment, and in the discursive framing of their desires (even the Civil Rights Era was not the start of 'new' collective action; see Kelley 2010). Indigenous peoples who challenged the colonial foundations of the nation-states who claim their territory were killed, forcibly relocated, and had their children removed from their homes to be culturally assimilated. These groups have been and continue to be dismissed as unnecessary and made invisible again, or rebuked for their radical, revolutionary, uncivil, and incipiently violent beliefs. What they wanted stood outside of the bounds of the political and social order, and was thus, impossible.

While so much of our political landscape has been shaped by the collapse of the Soviet Union, the end of the Cold War, and the largely unchallenged supremacy of the pax Americana, these interrelated phenomena do not explain all politics for the Left. For many communities, the end of the Cold War had no direct effect; there was neither an improvement nor a decline. Nor were their problems, concerns, and injustices suddenly born in the wake of the new world order. It is disputable that even separatism within the former USSR is 'new' to a post-Soviet context, given the sustained resistance and rebellion from several of its republics throughout its history (e.g. Chechnya). The diversity and history of hierarchy, discrimination, oppression and violence within nation-states were articulated and resisted. They are not new. These communities were not silent. We must not privilege the perspective of those who were not listening by arguing or even implying that these voices did not speak.

There is a 'citizenship gap' that is every bit as important as the 'rent gap' (Smith 1987). Whereas the rent gap points to the displacement of those whose rent is well below the perceived 'real' and higher rent a property could receive, the citizenship gap describes the displacement from the equal status the notion of 'citizenship' promises but does not, in practice, deliver. This gap is as old as the institution of citizenship itself, at least in Western democracies, for its institution was never what it claimed (Wood and Wortley 2010). In its specific allocation of rights, citizenship as an institution has served capitalist labour needs and liberal political needs (Marshall 1950). But its identity-based, uneven allocations were greeted with explicit articulations of the use of rights to create privilege, not equality and activist challenges to the citizenship gap. What Castells hears in the new social movements distinguishes them quite boldly from the struggles over identity politics on the (academic) Left:

> They [the new networked social movements] acknowledge the principles that ushered in the freedom revolutions of the Enlightenment, while pinpointing the continuous betrayal of those principles, starting with the original denial of full citizenship to women, minorities and colonized people. They emphasize the contradiction between a citizen-based democracy and a city for sale to the highest bidder.
>
> (Castells 2012: 246)

The 'post-political' describes an actually occurring set of political tactics, but *not a new one.* It is only new to those who believe (consciously or otherwise) in a universal experience, those who believe there was once a solidarity (usually rooted in organized labour) that has come apart. Those who see the post-political as new fail to appreciate the struggle and the discursive and physical violence that communities have endured. This view also mutes the fact that it was through political struggle that change was achieved, not gradual enlightenment. To call identity-based resistance 'new' erases that struggle, and elevates one political agenda over another.

Solidarity and intersectionality

When Fraser writes, 'In a sense, then, *we are flying blind*' (Fraser 1997a: 2, emphasis added), who is represented in the 'we'?[3] The starting place for her post-socialist Left does not take up Crenshaw and Collins' identification of the need, within social movements and among so-called allies, to recognize the internal jockeying for privilege and normalization of dominant experiences as representative. While diversity is acknowledged, its implications for the agenda and fundamental understanding of what constitutes politics are not engaged. There are presumptions by several scholars of solidarity past, present and future; the focus is more commonly on the imperative to build a coalition than on the complexity of doing so. Fraser is in favour of 'comprehensive, integrative, normative, programmatic thinking' and believes anything less is a confession of exhaustion on the part of the Left (1997a: 4). In fact, the fragmentation of and divisions within social movements are cited as motivations to move in the direction of unity, given the then-already evident consolidation and rise of the Right in American politics.

While supporting coalition-building, Mouffe argued strongly against false unity: 'it is important not to aim at a neutral conception of citizenship applicable to all members of the political community' (Mouffe 1993: 7). She seeks to incorporate pluralism within a radical understanding of democratic citizenship in practice, in which 'the many different struggles against subordination could find a space of inscription' under 'a common political identity'

(Mouffe 2005: 6–7). This identity is citizenship, but it is not apolitical nor undifferentiated. For Mouffe, antagonism will and must happen, for 'every identity is relational and that the condition of existence of every identity is the affirmation of a difference, the determination of an 'other' that is going to play the role of a "constitutive outside"' (Mouffe 2005: 2; Mouffe 2013). Mouffe refers to the work of Derrida, that 'the constitution of an identity is always based on excluding something and establishing a violent hierarchy between the resultant two poles – form/matter, essence/accident, black/white, man/woman, and so on' (Mouffe 2005: 141) However, while identity is relational and differential, it is not inherently hierarchical and violent, thus we must examine the circumstances in which it becomes so. I am a mother because I have children. It is a collective identity I share with others who identify as women and have had the experience of bearing and/or raising children (what it means to identify as a woman and as a mother is complicated, too: I know a trans woman who still goes by 'Dad' to her kids). The collective generation of meaning is important. Thus, I do not and cannot decide for myself and by myself what the identity of 'mother' means. As a collective identity, it may even mobilize for both literal and metaphorical purposes toward political goals, but it is not inherently hierarchical in relation to those who do not identify as mothers. Again, the circumstances that bring an identity, differentiated as it must be, into conflict with another to a point that it does produce hierarchy and violence must interest us. Mouffe's affirmation of difference and contest is welcome, but insufficiently historical for our purposes here. She writes of the need to avoid marginalization but presents too vague an understanding of how marginalization occurs, and how it is manifested in spatial patterns that reinforce it, and make it part of the social structure; addressing it is not simply a question of reversing and including.

Butler's contribution to the *New Left Review* discussion also intervened on this question, disputing the perception that 'we [the Left] have lost a set of common ideals and goals, a sense of common history, a common set of values, a common language and even an objective and universal mode of rationality…', as well as the belief that 'poststructuralism has thwarted Marxism…' (Butler 1998: 34). She warned that the 'unity' many sought was too often exclusionary and thus itself divisive, and challenged its proponents to consider their own role in these relationships:

> for a politics of 'inclusion' to mean something other than the redomestication and resubordination of such differences, it will have to develop a sense of alliance in the course of a new form of conflictual encounter. When new social movements are cast as so many 'particularisms' in search of an overarching universal, it will be necessary to ask how the rubric of a universal itself only became possible through the erasure of the prior workings of social power.
>
> (Butler 1998: 38)

The presumption of solidarity mutes and makes invisible the hegemony of white cis-hetero men's issues within the Left, and the consequent absence of others and of their experiences of discrimination and oppression along the axes identified by Young.

When Mouffe argues, 'It is always possible to distinguish between the just and the unjust, the legitimate and the illegitimate, but this can only be done from within a given tradition' (Mouffe 2005: 15), she makes an important point, but misses the opportunity to acknowledge there is no consensus around that distinction between 'the just and the unjust' – which is a judgement – and that there are and have been multiple, co-existing and conflicting traditions that frame such questions. There were opponents to slavery, to patriarchy, to colonialism; the dominance of those institutions was/is not merely a question of these voices being relatively silenced. When Spivak asks if those subaltern voices can speak and be heard, she exposes the power relationships involved in political participation, locked in the irons of legitimate and illegitimate subjects (Spivak 1988). What is also exposed (or should be) are alternate traditions, other senses of justice. The dominance for centuries of the view that slavery was acceptable and *justified* does not (or should not) erase the contemporaneous traditions of those whose sense of justice held that slavery was wrong. And I am not speaking, in the first instance, of white abolitionists, who broke from their 'own' traditions. I am speaking of those who were enslaved, whose suffering was monstrous and unspeakable, socially and legally. It was that suffering which rightly and appropriately grounded a politics in which challenging acts would produce conflict; the right and appropriate desire to be relieved of that suffering was the foundation of the idea of justice. It was not the sense of an abstract idea of fairness, or what it means to be human, or what government should look like. Nor did it produce a desire to have a civil, rational and unemotional debate about it, in order to resolve the issue.

There is a further problem in the terminology of 'cultural' which interferes with actual engagement with intersectionality, as its use seems to mask the social and physical violence that occurs beyond economic categories. Both Fraser and Young wrote (with differing degrees of criticism) about concerns that, at times, 'the politics of recognition is an end in itself' (Young 1997a: 156), without discussing issues for which it might be entirely appropriate: rape, laws criminalizing consensual sexual acts, laws criminalizing cultural practices, disproportionate rates of incarceration, injury and death at the hands of the police. Situating this violence within politics of recognition is necessary work to destabilize the idea that these are individual acts against individual people, and explicitly mark the collective identities that make individuals targets. At the scale of the collective, this violence is political and social, not merely psychological. Active recognition that the political and legal status quo is not neutral but is, rather, intentionally working to privilege some groups over others is still needed.

Moreover, neither allows for the possibility that separateness, even temporarily, might be an entirely legitimate and politically productive response to the experience of violence. They miss the politics, on many levels, of the trauma marginalized individuals and groups have suffered. As such, they also miss Crenshaw's arguments regarding the centrality of violence in the control of women – and experienced differently, for example, by white and Black women – and the need to see violence against women as 'social and systemic' and thus, political (Crenshaw 1991: 1241). As Catherine MacKinnon has written,

> No law silences women. This has not been necessary, for women are previously silenced in society – by sexual abuse, by not being heard, by not being believed, by poverty, by illiteracy, by a language that provides only unspeakable vocabulary for their most formative trauma, by a publishing industry that virtually guarantees that if they ever find a voice it leaves no trace in the world.
>
> (MacKinnon 1989: 239)

Ruth Lister admires Fraser's framing, but challenges Fraser's 'we' in a way that highlights embodiment. She turns the debate in a direction similar to Crenshaw's. She raises Fraser's silence on disability, and also asks, 'in a breathtakingly unequal globalizing condition, *who* do we recognize and *to whom* are we willing to redistribute?' (Lister 2007: 157). Lister also emphasizes the need to 'acknowledge the psychological pain' that accompanies 'misrecognition' without losing the material and political realities of their condition (166). Resistance to the silencing of trauma often requires reassertion of difference to expose the processes of systematic exclusion. The discussion of redistribution always takes place within a context of differentiated bodies who are allocated unequal value.

There is also some confusion and conflation within the discussion, regarding different types of identity. Class as a purely economic status – leaving aside for the moment the identities and cultural practices that arise from it – is fundamentally different from other identities. It is argued (fraudulently) that one's status is related to one's effort, that the hierarchy reflects what one 'deserves' by one's choices and actions, and the 'solution' to its inequality is to work harder and make sounder choices. But class is inherently unequal: it is, by definition, the unequal distribution of resources and labour. All other identities are not inherently unequal. The characteristics that are made to define the identity precede any action, even the individual existence of any person. They become hierarchies for the purpose of hierarchy, for the purpose of domination. Their difference is not inherently hierarchical; it is not based on uneven distribution. One cannot leave one identity for the other. They are used as code for class, sometimes, to justify further its hierarchy or as a 'modality in which class is lived' (see Butler 1998: 38). One can move from a working-class job to a middle-class manager. One cannot change one's gender, sexuality, disability, by working harder or knowing the right people. Acquiring the accreditations for class mobility does not erase the intersectionalities that distort meritocracy.

Fraser is sincere in her desire to bridge the two camps or 'paradigms of justice', but her understanding of politics reflects a privileged position, and misunderstands the politics of the marginalized. Although, in her response to Butler, she argues that misrecognition may be material, her discussion of 'cultural or symbolic' justice is fleshed out with examples that are communicative, rather than examples of their violent material expressions. Fraser's theorizing misses the part where groups are rejected and excluded, not just given less. And while she mentions lynching, her analysis of justice does not grapple with the violence that may explain why identifying with the 'cultural' group, as she argues, 'supplants class interest as the chief medium of political mobilization' (Fraser 1997a: 11). As Fraser reads it, identity-based discrimination displaces consideration of other forms, and thus these groups will identify 'the fundamental injustice' as cultural rather than economic exploitation. The consequent agenda for their political resistance will be to rectify the misrecognition, rather than fighting for redistribution. Both the mutual exclusivity of the two and the implicit superficiality of misrecognition are problematic. Mouffe has a related objection to any idea that identity is fixed or essentialist, that identity is formed in the process of mobilization and articulation of its target. I agree *in theory*, but the reality of the experience of these social relations, the violence with which they are enforced, is so very near essential that we must respect a politics that starts from a durability of identity that is experienced and enforced as essential. We should also recognize that 'essentialist' identities are commonly fixed in the first instance by governors and those who seek to exclude, not the governed or excluded.

The Left's anarchist politics have not been immune to these struggles. Graeber reports from his own experiences that when a movement had achieved some measure of success and had to determine next steps, at that moment it would become entangled in identity politics, which he felt was a displacement from the actual issue. In a footnote, he emphasizes that he is not dismissing questions of 'racism and privilege'.

> What I would argue is that the *way* that racial and class [sic] have been debated in the movement seem to have been startlingly ineffective in overcoming racial divisions in the movement, and I suspect this is at least partially because these debates are, in fact, veiled ways of arguing about something else.
>
> (Graeber 2011: 17)

It also raises the question (as I believe he intends) as to how these issues are discussed in and of themselves. If, when movements address them, they are really arguing about something else, then it indicates they have yet to engage such identity politics substantively (see also Morgan 2014). As Graeber also acknowledges, despite efforts to diversify (with some success; see Graeber 2013), anarchist activism remains dominated by white men.

It is worth briefly adding here that neoliberalism exploits its own identity politics: political parties that seek the privatization of health care, for example, also seek state intervention into the reproductive rights of women. No one is surprised by such lack of consistency or purity in politics, but what is worth underscoring here is that this neoliberal era treats different subjects differently. Heterosexual women's bodies are policed as heterosexual men's are not. Queer bodies are policed as heterosexual bodies are not. Black urban bodies are incarcerated for drug use while white suburban bodies are not. It is increasingly productive for the Left, as it theorizes and organizes its resistance, to read 'neoliberalism' not as an economic plan or a faith in the operation of markets, but as a broad, multi-headed programme to reassert white cis-male hegemony and redistribute not just capital but political and social power back into those hands (Hohle 2015; see also Graeber 2011; Harvey 2016a; Gough 2002). What drives decisions on the ground, as they happen? What drives individual moments of legislation, of interpersonal violence? If we believe it is only an economic logic that is ignorant or uncaring of its effects, then we repeat and retrench the idea that the sexism, racism, homophobia, ableism, and transphobia that accompany the material and spatial manifestations of neoliberalism are accidental or incidental. By the same logic, the distribution of suffrage to property-owning white men was similarly accidental, when we know it was not. In the face of articulations arguing the case for the actual, complete application of their lofty principles of equal citizenship, they chose not to do so.

If it is not an accident or an oversight, what is it? It is a strategic, intentional exclusion. What would be the rationale behind such exclusion and violence? It can only be a belief in the supremacy of themselves, and a desire to assert such a hierarchy through practices and spaces of exclusion. These practices have to be recognized for their intentionality. The continuum of domestic violence and street assault with legislative measures that disempower must be seen as a coherent discourse of oppression at all scales. Consequently, we must be extraordinarily cautious about how we imagine any role for the State in either redistributive or recognition remedies. The State has never been neutral. It has never been constituted as such, nor intended as such.

More, better democracy?

What, then, is a path forward for the marginalized in longstanding conditions of the post-political, where their activism and resistance is invisible to their allies and their goals deemed impossible? What does it mean to bring about change? What needs to be changed and to what form, to be deemed successful and transformative change, and not merely the exchange of one police force for another?

The proposed solution from 'the Left' is, largely, an increased radicalization of democracy. In the preface to the second edition of their 1985 work, Laclau and Mouffe argue for 'the necessity of redefining the project of the Left in terms of a "radicalisation" of democracy' (Laclau and Mouffe 2001: xv). As

Mouffe had previously advocated, through a radical embrace of pluralism under the umbrella of a citizenship based on political principles which in turn accept and embrace difference and conflict, we would intensify democracy without erasing difference.

Fraser also asked, what is democracy in the post-socialist era?

> Should we take it to mean free-market capitalism plus multiparty elections, as many former Cold Warriors now insist? Or should we understand democracy in the stronger sense of self-rule? And if so, does that mean that every distinct nationality should have its own sovereign state in an "ethnically cleansed" territory? Or does it rather mean a process of communication across differences, where citizens participate together in discussion and decision making to determine collectively the conditions of their lives?
>
> (Fraser 1997a: 173)

And what would any such a radical democracy require? Radical democracy, she argues, requires removing any obstacle that prevents full participation, including misrecognition. She acknowledges that when the debate appeared to be about to move forward with an intersectional approach (but stalled), she thought it would be a positive advancement. Fraser recognizes that it is possible to assert the category, for example, of 'women', without losing sight of how the definition of the category is malleable. She therefore argues for the need to place categorical definitions and the assumptions that drive them in their cultural and historical contexts (Fraser 1997a). But while her idea of contextualization means 'specific to time, place and culture', context should also include the ways in which identities are historically as well as socially produced and come to signify different things.

Nevertheless, she continues to see a problematic bias in favour of cultural recognition at the expense of redistribution, and finds recognition politics mired in conflicts 'which rage across the whole spectrum of "new social movements." These arguments pit antiessentialists...against multiculturalists.... The issue at bottom is the politics of recognition: *Which* politics of recognition best serves the victims of misrecognition? Revaluation of difference or deconstruction of identity?' (Fraser 1997a: 174). To my mind, a more important concern is who is *asking* these questions and to what end? Fraser's concern with doing things the 'right' way is quite different than the 'by whatever means available' that is often the reality for marginalized groups, given the way they inhabit the city and their allotted place in the public sphere. In a desire for unity on the Left, Fraser focuses on tensions among progressive approaches and fears they are or will become incompatible, rather than viewing them as making different contributions.

There are parallel weaknesses in critical urban theory as it searches to define a radical politics of the city. Literature with an emphasis on a narrowly framed political economy and redistributive politics in many instances does not take up the politics of the diversity of the lived social geography of cities. Brenner, Marcuse and Mayer argue that it is 'urgent' to address the question of 'how this crisis has provoked or constrained alternative visions of urban life that point beyond capitalism' and the need for 'critical or "revolutionary" theory' in 'charting the path' to another kind of city (Brenner et al. 2011: 3). Among the concerns of critical urban theory, they list the intent 'to expose the marginalizations, exclusions, and injustices (whether of class, ethnicity, 'race,' gender, sexuality, nationality, or otherwise) that are inscribed and naturalized within existing urban configurations' and 'to demarcate and politicize the strategically essential possibilities for more progressive, socially just, emancipatory, and sustainable forms of urban life' (Brenner et al. 2011: 5). However, the realities for daily life and politics that arise from those 'marginalizations, exclusions, and injustices' are missing from their discussion. Grounding our inquiries in the inhabited city leads us to see what has been missing from this radical political thinking – not what is wrong, but what is incomplete.

For critical urban theory to draw on these radical politics, the role of the urbanization of the population and the identities and spaces it creates for politics – both oppression and resistance – must be incorporated. Keeping in mind that urbanization is occurring at different times and on different time scales in different places, it is a process that fuels the processes of identification and antagonistic relationships that Mouffe and Isin discuss. New populations will come into the city, inhabit it differently, and challenge and reinvent its practices. They will also come up against the wall of elite power, claims of unity, and 'higher' forms of civic identity that actually exclude. In other words, the chronology of critical urban theory and politics should map onto specific processes of urbanization, as Purcell has indicated, not (only) geopolitics. This perhaps seems obvious, but it is not generally what we have done.

Purcell proposes that the road to radical democracy should be explicitly grounded in the city. He acknowledges that 'we live in a world that equates democracy with a liberal-democratic state, which is a form of oligarchy that sets severe limits on democracy and insists that anything beyond those limits is impossible' (Purcell 2013: 26). Democracy, for Purcell, is not the State. As he elaborates:

> Keynesianism imagines the state to be essentially the same thing as the public, in the sense that the state stands in for, and acts as if it were, the people. But in fact, the state is not the people. Nietzsche's Zarathustra had it right: "State is the name of the coldest of all cold monsters. Coldly it tells lies too; and this lie crawls from its mouth: 'I, the state, am the people.'" The state is only ever a very small subset of the people. Democracy means that the people rule themselves.
>
> (Purcell 2013: 10)

Purcell argues that we should pursue 'active democratic autonomy'. In his eyes, it is basically a life force: 'Democracy already lives in the body of our current society. It produces itself' (Purcell 2013: 25, 27). He advocates for a Lefebvrian approach that begins with a grounding in the concrete life of the city and moves towards a horizon, never arriving, of self-rule. Lefebvre calls this process 'transduction'. Imagining and then creating, and further re-imagining alternative spaces and practices occurs when individuals and communities begin 'to "step back from the real,' to refuse to accept what already is, but also to never lose sight of that real, to always begin from the activity people are already engaging in' (Purcell 2013: 25).

'Transduction is a project we all take up together' (Purcell 2013: 27). Here is where Purcell's proposals would be enriched by more specific detail of the inhabited city, and an engagement with intersectionality. Transduction is something we can (and do) all take up, yes, and it resonates strongly with political and social strategies of resistance in marginalized communities that I will address in the following chapter. But it is not, at least currently, something we can take up *together*. It is important to recognize that the city is an antagonistic, not united, space, and such strategies of expanding autonomy are taken up precisely in opposition to, or in evasion of, the unifying efforts of the State. The twenty-first-century city does not provide, nor it is likely to, a social unity conducive to the formation of stable solidarities that this approach requires.

I argue that the grounding of radical democracy as proposed by critics of the post-political is insufficient, and it does not address the theoretical and practical challenges of intersectionality. These scholars remain focused on coalition building without addressing its actual viability or interrogating their own intersectional positions. Purcell is moving in a more productive direction with his groundedness in Lefebvre's *l'inhabiter* and process of transduction. Nevertheless, radicalization of democracy as a political agenda does not quite come to terms with the intentionality of the violence and exclusion, nor with Crenshaw's identification of jockeying for position that happens within society and within the Left.

In some ways, the 'post-political' is an unfortunate term for these purposes, because it suggests we have moved past an era – which suggests that there *was* such an era – where there was an alleged openness to discussion and difference of opinion. The post-political critique's argument for exposing the strategies and tactics of the foreclosure of possibilities remains salient. The point is that this has been a strategy against challenges to the status quo for centuries.

We must expand temporally the analytical reach of the post-political, and release it from the specifics of the post-Cold War era. The challenge is not merely to re-categorize our historical eras, but to allow for multiple, co-existing chronologies. Perhaps it would also be productive to treat the post-political not as an era, but as a condition that may be attempted in many circumstances and contexts. To do this, we must also recognize the privileged position of its critics, in their assumptions that their experience was new. Even these critical (and empathetic) thinkers have treated the experiences and geographies of which they had little or no first-hand knowledge as invisible. And their idea of politics was still the ability to enter into debate, to have the other side (the elite) acknowledge the possibility of a difference of opinion. For it was a perception of a similar loss – their own – that produced the observation of strategic 'foreclosure'.

What are the processes, the political strategies and tactics, the acts of citizenship that seek to bring about the change as defined above, and what do they tell us about what it means to be political? How can our understanding of these acts redefine the political in a more radical way towards the advancement of critical urban theory? Mouffe, Fraser and Purcell's moves towards more inclusive, radical democracy are welcome, but if we look at the inhabited city, we can see that their chronological frameworks are limited or problematic, their politics are less than comprehensive, and they have not fully engaged the actual politics arising from intersectionality, antagonism, intentional violence, and embodiment that Young, Crenshaw, Collins and Lister have drawn to our attention. Fuller engagement, I argue, would compel the Left to shift its political frame in theory and in practice. The final two chapters attempt to grapple with these particulars.

Notes

1 Reproduced by kind permission of the Estate of Sean O'Casey.
2 The US Department of Justice subsequently filed civil rights charges against the officers, two of whom were convicted and served prison time.
3 In fairness, Fraser elsewhere puts 'we' in quotes; see 1997a: 218.

3 Sad, sick and diva citizens

Resistance, refusal and urban space

In this chapter, I want to recognize and attempt to articulate a politics that starts not only from thought, but from suffering and pain. As a woman in Black Lives Matter – Toronto put it, 'When you dig into it, at the root of it all is pain. As black people, how does it make us feel to shout out over and over that our lives matter? Do I even have to say that? Sometimes when I scream it out, I say it to myself, to remind myself' (Battersby 2016).

Many theories of resistance, citizenship and justice start from positions of power, rather than positions of suffering. These are often articulated, knowingly or not, through a politics of respectability frame: the idea that democracy is a fundamentally civil conversation, even if spirited or passionate – and this is the case in representative democratic institutions *and* in activist consensus circles. This democracy of the civil conversation preserves power relationships. 'In the patriotically-permeated pseudopublic sphere of the present tense, national politics does not involve starting with a view of the nation as a space of struggle violently separated by racial, sexual and economic inequalities that cut across every imaginable kind of social location' (Berlant 1997: 4). But it should.

How can politics emerge from suffering? What does a politics look like that is equal parts visceral and rational? How does it express itself? What can it achieve? How can we connect it to other understandings and acts of politics? This politics is premised on the knowledge that starts from and is acquired through suffering. It aspires to justice and to citizenship, and that aspiration is productive of definitions of both. Cooper referred to these forms of knowledge as 'epistemologies from the margins' (Cooper 2014). The earlier discussion of strengths and limits of the post-political literature reminds us of the need to examine the experience and politics of those who are excluded from the definition of the citizen and the national body, those whose suffering oils the gears for the power and comfort of others, and whose suffering is intimately, bodily and directly related to the production of the nation and of national identity (see Cohen 2004; Berlant 1997). The study of politics must be explicitly intersectional and its temporal and spatial frames must reflect that. This argument will situate acts of resistance in their neoliberal urbanist context, but will also map them against geographies of crisis that differ from those of dominant culture and/or global political economy approaches.

For the mainstream and even for allies, this is a suffering that is invisible, and a consequent politics that seeks (what we are told is) the impossible. Expectations as to what form its politics will take need to expand. It is important to remember 'the simple fact that most subordinate classes throughout most of history have rarely been afforded the luxury of open, organized, political activity. Or, better stated, such activity was dangerous, if not suicidal' (Scott 1985: xv). Their acts are indicative of both courage and desperation. Often they reflect a need to scream in order to be heard, and how bad things have to be before it feels worth the energy to scream.

Recognizing resistance in its queering of space and behaviour (Berlant 1997), in its foot-dragging (Cohen 2004) and in its refusal to engage (Cohen 2010; Simpson 2014) continues to widen the lens of what constitutes acts of citizenship (Isin and Nielsen 2008). Kalyvas has reminded us that in the study of democracy we too often ignore acts that fall outside of formal political spaces, and place most of our focus on 'normal politics and ordinary lawmaking – that is, on instituted democracy' (Kalyvas 2008: 295). Drawing on the idea of 'extraordinary' politics, as articulated by Max Weber, Carl Schmitt and Hannah Arendt, Kalyvas argues they 'open up imaginative ways to think about the significance of unpredictable and discontinuous deeds that defy the established order, challenge the scope and content of institutionalized politics, and transgress the limits of the possible and the accepted' (Kalyvas 2008: 292).

> The extraordinary is a reminder that instituted reality does not exhaust and cannot consume all forms of political action, which often emerge at the edges of the existing statist *nomos*.... From this point of view, phenomena such as civil disobedience, irregular and informal movements, counter-institutions, protests, insurgencies, street fighting, and illegal upheavals are as (if not more) important to democracy as normal politics.
>
> (Kalyvas 2008: 297)

Kalyvas' focus is on collective action and public acts. Literature on social movements and resistance, particularly in Western countries, has begun to challenge the underestimation or dismissal of collective action that happens without formal organizations (Cox and Nilsen 2014; Runciman 2016; Véron 2016). Drawing on the work of Cathy Cohen, Lauren Berlant, Johanna Hedva, Audrey Wollen and Davina Cooper, I want to build on Kalyvas' idea and apply its insights to other forms of resistance that are smaller, less visible, and maybe not even collective as such, but which are just as 'extraordinary'. Although still with a focus on collective, public action, Harrison and Risager encourage us to 'think about organization in a way that includes both the stable and the fluid, the formal and the ephemeral, the rational, conscious decisions and confusion and affect' (Harrison and Risager 2016: 846). Many of the acts I will discuss here are informal, temporary, even ephemeral. They are a result of, and productive of, the nature of the politics. They include banal, everyday spaces and acts, sometimes for extraordinary purposes for publicity, sometimes to remain intentionally unseen. The way in which city space is used is critical to the form and meaning of these politics.

The central purpose here is not to theorize the nature of suffering to any great extent, but to think about what kind of politics might arise from suffering and societal abjection, how we might recognize it and understand it, and what type of citizenship is constituted through such political understandings and acts. Placing many post-crisis activisms in this context, of the politics of the marginalized in their everyday life in the city, suggests another way to explore the horizontalism, the importance of symbolism, the lack of legislative demands, and the emotion even of large, organized movements such as Black Lives Matter and Idle No More. This is not a specific consideration or analysis of their strategies, but to suggest they are shaped by other political genealogies and other crises, in addition to the global financial crisis and its aftermath of austerity.

Consider the following events:

In October 1988, in front of a federal building in Los Angeles with 40 fellow protesters, Paul Longmore, who could not use his hands due to having polio as a child, burned his newly published first book, the product of ten years of study and writing, to protest the classification of its royalties as unearned income that thus jeopardized his eligibility to work and receive government assistance (Longmore 2003).

In 2002, for the first Talking Stick Festival in Vancouver, Anishnaabe artist Rebecca Belmore stood on a street corner in the city's Downtown Eastside, scrubbed the sidewalk as she might do for a kitchen floor, wrote the names of murdered and missing indigenous women on her arm, screamed their names, dragged a rose through her teeth after each name, and then put on a red dress and nailed it to a post behind her. She then tore the dress away from the nails, until it was in shreds and she stood in her underwear (Balzer 2014).

In November 2006, on Albanian Flag Day, members of the Albanian nationalist movement, Vetëvendosje, threw bottles of red paint against the walls of the government building and the UNMIK [UN Mission in Kosovo] in Prishtina, Kosovo, in protest of the negotiations. The red paint was a symbol of the Albanian flag and of blood. Some protesters also threw stones. The police responded with tear gas. Vetëvendosje's campaign, by their own account, included 'writing graffiti on walls, on UNMIK jeeps, deflating their tires, throwing paint at buildings, throwing eggs, boycotting Serbian products, blaring an emergency announcement at UNMIK and blockading institutions and the border' (UN Human Rights Advisory Panel 2015).

In April 2010, members of the Grassy Narrows First Nation of Ontario marched on the provincial legislature carrying a full kilometre of blue fabric to represent 'a wild river', in protest of ongoing problems resulting from mercury poisoning of a local river in the 1960s. In June 2016, protesters poured out grey liquid from four barrels labelled with a skull and crossbones and 'mercury kills' in front of the Ontario legislature (Leslie 2016). Six protesters were arrested and charged.

In September 2010, about 100 Roma stood outside the French Embassy in Bucharest, with pictures of large fingerprints held in front of their faces, to protest mass expulsions of Roma from France to Bulgaria and Romania (Council of Europe 2013).

In December 2014, Kasha Jacqueline Nabagesera launched *Bombastic*, a magazine by and for the LGBTI community in Uganda, where homosexuality is illegal, and where politicians and media have harassed individuals by name and called for their deaths (Fallon 2015).

In May 2016, disabled persons in Cochabamba and LaPaz, Bolivia suspended themselves with ropes in their wheelchairs from a freeway overpass, and blocked street traffic by lying in a row across on their backs in their wheelchairs, as part of a campaign for a significant increase in support payments for disabled persons. In previous protests, police had responded with tear gas (Ly 2016).

In July 2016, Dalits in Gujarat stopped performing their usual skinning and scavenging of dead cows, work for which they are routinely harassed and assaulted. Protesters left rotting carcasses in the street and some also hauled them to government offices (Menon 2016).

These are acts of citizenship of a different register. The participants often make themselves more vulnerable as they put their vulnerability on full display, thus making that vulnerability dangerous, as well as in-your-face. Many of the acts are spectacles with no obvious goal, or immediate hope of achieving one. Organized or spontaneous, the protests are more notable for their emotional aspects – either the emotions of the participants or the obvious desire to provoke an emotional response from observers. These kinds of acts are often dismissed in some quarters, in much the same way as Occupy was. Critics, including fellow activists, see them as disorganized, without agenda and unserious: pointless drama.

Marginalization and suffering in the city

In order to make sense of the politics of the above acts, the nature of the marginalized's inhabitance of the city must be explored. Thus, the specific relationship to the city and urban life needs to be highlighted as a complement and counter to the limits of political economy approaches to the city. Thinking through some of the details of histories of the politics and activism of marginalized groups, and their use of the city in emancipatory and explicitly political ways, serves as a window into understanding today's protest on its own terms, and understanding the richness of politics in the city.

As bell hooks wrote, 'Marginality [is] much more than a site of deprivation.... It is also the site of radical possibility...for the production of counter-hegemonic discourse that is not just found in words but in habits of being and the way one lives....It offers...the possibility of radical perspective from which to see and create, to imagine alternatives, new worlds' (hooks 1991: 149–150, cited in Cooper 2014: 32). Cooper notes feminist and other critical theory has insisted on exposing the relationship between power and the production of knowledge, and highlighted the need to situate analysis and to refuse ideas of objective or universal truths. With her idea of 'epistemology from the margins', she argues for

> reading social life through and from *non-dominant* experiences and forms: for instance, understanding the body through corporeal forms that are not hegemonic but female, disabled, and queer (or as viruses, transplants, or surgery); understanding property not through private ownership but through squatting, common or public lands. Epistemologies of the margins are not simply intended as perspectives from which to critique mainstream, hegemonic forms...
>
> (Cooper 2014: 32)

'Epistemology from the margins' ought to shift our perspective in a transformative way, and create academic and political space for the knowledge of people whose centre is the dominant's margin (Wood 2001).

But let us first pause to reflect on the city as a 'difference machine' (Isin 2002) and an active producer of marginalization. The city itself and its urbanization processes generate the suffering and institutionalize it in the built landscape and social space of the city. For example, critical disability studies emphasizes the *enabling* of certain bodies by the built and social environments. Disability is 'an elastic and dynamic social category. It is not an objective condition. It is a set of socially produced, highly mutable, historically evolving social identities and roles' (Longmore 2003: 239). City form is thus productive of 'disability'. The city is socially and physically fragmented to privilege one group over another (Amit-Talai and Lustiger-Thaler 1994; Graham and Marvin 2001). City space is structured by law-makers, property owners, and capitalists to regulate and restrict movement and activities, and is policed and governed so as to enforce that order. For many individuals and communities – those in possession of unwanted or expendable bodies – the city as they inhabit it is a place where survival is a daily concern. Pressure comes from outside and inside communities within a landscape of moral panic, and the different types of 'policing' reflect the many paths of governance. The pressure is ever downward: 'in this postindustrial era we face advanced marginalization wherein members of marginal communities have taken on the daily, face-to-face responsibility of policing the individuals in their group who have less resources and power' (Cohen 2010: 42).

The spaces of abjection within cities brought about through redlining, poverty, racialization, migration practices, lack of access for those with disabilities, exposure to environmental hazards, and differentiated access to resources and infrastructure, make identity differences visible and material through their spatial orders. 'Poor (largely ethnic minority) communities were frequently dispersed, and interstate highways were built through densely populated urban areas, which disrupted communities and further supported the migration of (mainly white) middle class and affluent populations away from the city centre' (Bofkin 2014: 12). Travellers and Roma have frequently been resettled away from both city centres and rural areas to poorly serviced urban margins (Crowley 2009; Wood 2006). Although its manifestations are particular to its contexts, the segregation of immigrant neighbourhoods and racialized communities persists in Western cities worldwide, and by some measures has intensified since the 1990s (Iceland 2009; Massey and Denton 1993; Arbaci and Malheiros 2010; Van Kempen and Murie 2009). These same spaces may in turn create the autonomy, safety and solidarity necessary not just for action, but for healing. Importantly, however, they are regularly perceived as both a danger and an obstacle to the alleged unity of the nation or the city. For example, in her work in the UK, Cooper found a mainstream resistance to the exclusionary spaces its normative practices had created. She heard such neighbourhoods described as 'ghettos', a kind of societal sickness caused by groups unwilling to restrict their cultural differences to their homes, and thus a threat to unity and 'a common citizenship' (Cooper 1998: 134).

Yet the city also creates the opportunity and the stage for disruption to a highly public audience who does not easily have the choice to look away. Indeed, that public is habituated (though not inured) to such interruptions and considers them as part of the urban landscape (though sometimes unwanted). The city's mobility, fluidity and publicly accessible spaces have long created possibilities for new social freedoms for oppressed and marginalized groups (Stansell 1987; Strange 1995; Chauncey 1995), as well as formal and informal territorializations (Gieseking 2016), that in turn enable possibilities for dramatic disruption.

When we consider the daily struggles and suffering, one key aspect to highlight and integrate into our understanding of urban politics is that the citizen who inhabits this city is an embodied, emotional subject. Castells has noted this in his review of post-crisis movements: 'At the individual level, social movements are emotional movements. Insurgency does not start with a program or political strategy' (Castells 2012: 13). For Castells, fear and enthusiasm are the key motivators of social movements, anger/outrage is a means of managing or overcoming fear, anxiety and despair, trust is foundational for any productive human contact, and enthusiasm is further energized by hope. Although infrequently, Castells also employs a vocabulary of 'suffering', such as when he writes of 'the stern critique of a merciless economic system that feeds the computerized Automaton of speculative financial markets with the human flesh of daily suffering' (245).

Castells suggests that 'the key question to understand is when and how and why one person or one thousand persons decide, individually, to do something that they are repeatedly warned not to do because they will be punished'

(Castells 2012: 13). While he is not wrong, it is productive to turn the question around, as Berlant does when she asks,

> How can U.S. 'minorities' abide, just one more day, the ease with which their bodies – their social labour and their sexuality – are exploited, violated, and saturated by normalizing law, capitalist prerogative, and official national culture?…What are the *technologies of patience* that enable subaltern people to seem to consent to, or take responsibility for, their painful contexts?
>
> (Berlant 1997: 222)

Berlant writes of the angrier, non-compliant corners of what she terms 'queer nationality'. The emotion and explosion of patience snapping are related, as well as critical and definitive; they compel and enable queers to push past the usual bounds of political and social safety. Referring to alternative zines, she argues, 'But they are also magazines in the military sense, storehouses for the explosives that will shatter the categories and the time-honored political strategies through which queers have protected themselves' (Berlant 1997: 170). This is an emotional state and a politics born of a daily, constant build-up of slights, micro-aggressions, exclusions, traumas, experiences of hate and of violence (Rankine 2015; Emmanuele 2015).

Emotions are essential to politics, not only as mobilizers as per Castells, but as subjects of contestation themselves. 'Feminist politics of emotions recognize emotions not only as a site of social control, but of political resistance' (Boler 1999: 112–113). The systematic reproduction of national identities in normative modes that exclude and shame results in an insidious and harmful manipulation of others' emotions. With reference to Sandra Bartky's work, Boler argues,

> It is itself psychologically oppressive both to believe and at the same time not believe that one is inferior – in other words, to believe a contradiction […]. She [Bartky] analyses specifically how women are enculturated to internalize a sense of shame and lack of self-worth as an oppressive effect of ideological contradictions. In short, many Marxist explanations of ideology have not systematically excavated the gendered dimensions of consciousness, nor emotions as a site of hegemony.
>
> (Boler 1999: 129)

Emotions are themselves public practices as well as feelings, shaped by their cultural contexts. They are used to govern citizens' behaviour, to create legitimate and illegitimate political subjects, and especially to compel citizens to govern themselves accordingly (Ahmed 2004). The spatial segregation and regulation of bodies and activities in the city institutionalizes and normalizes the abjection. From these experiences, those who are marginalized *know* that the daily functioning of the city is discriminatory and violent, and know that the State is at best an imperfect defender of their rights and at worst, complicit in the violence. They know that the emotions for which they are judged as irrational or 'out of control' and through which their oppression has been alleged to be justified, hold truths about what constitutes injustice (Kelley 2014). The dominant dissociate gender, sexuality and other marginalized identities and practices – especially their spatialities – from the public sphere, and thus, separate the emotional and the embodied from the 'rational' and determine the latter as the only mode fit for (acceptable) political acts. But those who are marginalized know that their suffering is political, that it is central to the economy and culture of those who claim to govern them. Engagement with the inhabited city brings this knowledge to the surface, which is not addressed in critical urban theory that is too exclusively anchored in a Marxian or political economy framework (see also Rose 1993; Bain 2004).

Diva citizenship, utopian spaces and the politics of refusal

There is no one single response, or type of politics, or political belief, that unites the diversity of issues, forms of resistance, strategies and tactics of the many groups discussed here. This discussion is not to suggest they are the same, or even unified or solidaristic. But there are some common practices that seem directly related to the nature of their situation, and these reveal a political sensibility rarely included in scholarly approaches to activist politics and critical urban theory. Many of these fall in my category of 'intransitive citizenship', in that they do not have a clear object or target. They are nevertheless acts of citizenship, and often powerful and effective.

Starting with Berlant's concept of 'diva citizenship', I argue for taking seriously activism of the excluded that is informal and unconventional. Similarly, the spaces it selects for its public engagement lie outside formal institutions and 'acceptable' places for political statement. In her theory of 'diva citizenship', Berlant has written of how and when the *private* suffering of individuals is excluded from 'the public interest' and thus politics, diva citizenship emerges through outrageous acts which put privately suffering and imperfect bodies in public settings where the act of removing them (which is a certainty) will expose the hypocrisy and the suffering. It does not achieve anything concrete in and of itself, but it (hopefully) stretches the frame of what is sayable and possible. Acts of diva citizenship have a particular interest in confronting the public with that which has been hidden in private, made invisible, for the discomfort it will cause those who are not afflicted. In the moment when it feels as though all the rules that seemed fixed are broken and thus suspended, a creative space appears in which a new thought, a new word, a new path is possible. Although Berlant does not explicitly describe it as such, diva citizenship through its quest for visibility is intensely urban by nature, wherever its acts are located: '[t]he centrality of publicity to Diva Citizenship cannot be underestimated, for it tends to emerge in moments of such extraordinary political paralysis that acts of language can feel like explosives that shake the ground of collective existence' (Berlant 1997: 223).

Berlant borrows 'diva' from bell hooks' idea of diva speech, where hooks 'argues that the transgression of a gentility that pretends social relations are personal and private rather than public, political and class-based must be central to any liberation politics', and from Koestenbaum's advocacy for 'performing the public grandiosity of survival and the bitter banality of negotiating everyday life in a way that creates a transgressive and threatening counterculture' (Berlant 1997: 224). Berlant's work also redirects us to examples of a deeper history of racism and homophobia, whose consequences were corporal, material and discursive, whose injustice was profound and devastating, and whose political context left little material or discursive space to occupy or speak. Acts of diva citizenship often lack a specific agenda or measurable goal; they are intended to expose the citizenship gap and 'shock white citizens into knowing how compromised citizenship has been as a category of experience and fantasy…' (238). Through this idea Berlant constructs a feminist, queer and anti-racist theory of the State and of law, and of the political. When, as she argues, the 'national symbolic order' into which you are born designates your body as inherently unworthy of citizenship, and your body itself is targeted for personal violence that is nominally illegal yet is fetishized and goes unpunished, the spaces for formal political action are few and far between. Political space must be blown open.

Spontaneous explosions sometimes bring results. Riots in 1930s Harlem brought about more substantive change to its residents from the Mayor's office than any effort to work through official channels (Greenberg 1992). But we must not see these explosions as callous or cavalier acts, nor miss entirely the pain at the root of any of these actions. When Paul Longmore burned his newly published book, he described it as 'a moment of agony', and one of his fellow participants believed that 'the entire protest, especially the burning of the book, gave tangible form to the pain they felt about their own lives' (Longmore 2003: 253). As his book disappeared in the flames, Longmore read out a statement, which included 'The burning of my book symbolizes what the government does to us and our talents and our efforts. It repeated turns our dreams to ashes. We find that outrageous, and we will no longer quietly endure that outrage' (258).

At the less visible end of the spectrum of political acts is Cohen's idea of 'deviance as resistance', which is the basis for a political theory 'that is centered around the experiences of those who stand on the (out)side of state-sanctioned, normalized White, middle- and upper-class, male heterosexuality' (Cohen 2004: 29). Grappling with the enormity and complexity of experience and perspective this entails, she makes the crucial point that,

> In contrast to many privileged gay, lesbian, and queer folks, poor single Black women with children, structurally unable to control an exclusive 'ghetto' or area of the city where *their dealings with the state are often*

> *chosen* and from an empowered position, are reminded daily of their distance from the promise of full citizenship.
>
> (Cohen 2004: 29, emphasis added)

From this relatively powerless position, individuals engage in 'counter normative behaviors'. These 'deviant practices', argues Cohen, are a form of resistance and a possible space for the emergence of 'a new radical politics of deviance'. Whereas academic studies of activism have frequently posited only 'structured, coordinated, and seemingly purposeful acts' as legitimate and serious, with a focus on leaders and formal organizations, Cohen builds on Robin Kelley's 'politics from below', and seeks to restore to our analysis 'the agency and actions of those under surveillance, those being policed, those engaged in disrespectable behavior'. These people, too, are political subjects. Although their choices 'are by no means chosen freely in the liberal sense', we should nevertheless recognize their 'struggle…to maintain or regain some agency in their lives as they try to secure such human rewards as pleasure, fun, and autonomy' (Cohen 2004: 38).

The Los Angeles artist Audrey Wollen's 'Sad Girl Theory' posits suffering as resistance and argues that we should recognize its invisible acts as politics. For Wollen, turning intense emotions inward, although 'devastating', is also a refusal to ignore or pretend to transcend the harm and violence of daily life and extraordinary trauma (Wollen, quoted in Salek 2014). Wollen challenges the exclusive focus on public, often violent acts of resistance as legitimate protest; even peaceful protest is only considered seriously when in the form of a public mass mobilization:

> I'm trying to reveal how narrow that definition of activism is, that there was actually a whole lineage of women who resisted the oppressive structures through what has been incorrectly defined as passivity. Sorrow, weeping, starvation, and eventually suicide have been dismissed as symptoms of mental illness or even pure narcissism for girls. I'm proposing that they are actually active, autonomous, and political as well as devastating.
>
> (quoted in Salek 2014)

Joanna Hedva's adaption of 'Sad Girl Theory' into 'Sick Woman Theory' carries the argument further still. She reminds us that the vulnerability of the body is incontrovertible, and our suffering is the product of and often the fuel for the system we resist. Those who inhabit excluded and exploited bodies know their excluders and exploiters would destroy them. Hedva thus concludes, 'The most anti-capitalist protest is to care for another and to care for yourself' (Hedva 2015), echoing Audre Lorde's statement, 'Caring for myself is not self-indulgence, it is self-preservation, and that is an act of political warfare' (Lorde 1988: 131). As Lorde also wrote, when she was facing a terminal diagnosis of cancer,

> In this loneliest of places, I examine every decision I make within the light of what I've learned about myself and that self-destructiveness implanted inside of me by racism and sexism and the circumstances of my life as a Black woman. … Survival isn't some theory operating in a vacuum. It's a matter of my everyday living and making decisions.
>
> (Lorde 1988: 60)

Hedva's Sick Woman Theory concurs with Wollen 'that most modes of political protest are internalized, lived, embodied, suffering, and no doubt invisible'. We cannot help but react and be affected by 'regimes of oppression – particularly our current regime of neoliberal, white-supremacist, imperial-capitalist, cis-hetero-patriarchy. It is that all of our bodies and minds carry the historical trauma of this, that it is *the world itself* that is making and keeping us sick' (Hedva 2015; see also Emmanuele 2015). Due to this vulnerability, she argues, our politics must concern the need to structure social, economic and political systems around its protection. We must never lose sight of the mental and physical well-being of the body itself, and must dispute the manipulation of the body's vulnerability to misrepresent societal failure as individual weakness.

Berlant agrees here, too:

> Whenever citizenship comes to look like a question of the *body*, a number of processes are being hidden. The body's seeming obviousness distracts attention from the ways it organizes meaning, and diverts the critical gaze from publicity's role in the formation of the taxonomies that construct bodies publicly.
>
> (Berlant 1997: 36; see also Yingling 1997)

Berlant's 'diva citizenship' arose from the experience of being excluded, but also exploited, emphasizing that the labour of abject bodies was fundamental to the identity, productivity and wealth of those in power who then refused to acknowledge commonalities with, violence done to, or even the existence of those abject bodies.

The daily suffering is alleviated (sometimes) by what Davina Cooper calls 'everyday utopias – networks and spaces that perform regular daily life, in the global North, in a radically different fashion' (Cooper 2014: 22). Although 'utopia' also holds a connotation of the fantastical and of unachievable perfection, Cooper's everyday utopias are grounded in the real world. They are not removed from daily 'mainstream life', but rather constitute a response to it and new spaces and practices within it. This turns our attention to the capacities of informal practices and spaces to satisfy needs and desires that *cannot* be replaced by the formal. Cooper describes it as 'a pragmatism oriented to survival' and acknowledges along with others 'the dynamic, improvised, often flawed quality of many utopian spaces' (7). Rather than working in more formal, organized ways, with a goal of bringing about legislative change, change in political office or even taking office, utopian activists seek to create change for themselves, and reclaim or invent the spaces in which to do so. The desire to generate these new spaces is directly related to the alleviation of suffering, and there is often a complicated relationship to history and past trauma. In many instances, to forget the past is both a further trauma and a danger, even as there remains a hope to transcend its pain. Neither is the past a singular or simple thing; in fact, 'the past is not always identified as what is wrong; indeed it may also be what needs recuperating' (Cooper 2014: 222–223).

De-centering the dominant perspective includes refusal, as well as direct challenge. Everyday utopias are a form of refusal, of choosing not to accept the status quo and instead carving out space for something better. The politics of refusal of indigenous peoples reminds us that so-called 'alternative' paths are not always in response to the dominant, but may exist in their own right or even pre-exist it (King 1990; Simpson 2014). Refusal and an assertion of parallel knowledge systems are found throughout Aboriginal scholarship. Thomas King (1990) has argued for a refusal of applying the 'post-colonial' frame to Native creative writers, because its historical frame is inherently Eurocentric and is irrelevant to the development of indigenous storytelling. Craig Womack (1999) and Jase Weaver, Womack and Robert Warrior (2006) have similarly advocated for Native literature to be understood as its own canon on its own merits, with its own traditions. Indigenous writers have refused Western genres and incorporated oral storytelling and their understandings of what constitutes a story into their written work (King 2003; Cole 2006). Many of these scholars have also refused a separation of processes of academic and literary categorizations from politics, recognizing these as issues of sovereignty. Those explicitly studying politics have sought to reclaim their own political practices (often proscribed through settler laws) and reassert sovereignty, which begins with a refusal of settler geographies and territorial claims, as well as the cultural violence that accompanies them (Alfred 1995; Alfred 2005; Simpson 2014).

Autonomy and internal logic form part of Cooper's utopianism, too. Cooper argues that a community's reinventions and transformations are rooted in more than direct opposition, but are products of independent desire: 'Utopian concepts do not derive simply and only from the world they reject; they are also forged with some degree of autonomy through their anchorage in the utopian world that is grown' (Cooper 2014: 33). But autonomy is complicated: Berlant speaks cautiously and wisely about the marginalized (she refers specifically to queer subjects at the point in question but it applies more broadly) and their inability to fully exile themselves from national identity, from the State and its support and protection – even though these are a double-edged sword, with the State's assimilationist policies and policing.

In the idea of 'diva citizenship', citizenship remains a contested relationship with those who govern, both legally and socially. Tellingly, for occupation movements, there was at best a tolerance, and commonly a ferocious police response to clear encampments and restore 'normal' purposes of urban space. Forcing the governor to respond and to reassert control, when the absurdity of the inequity or oppression has been made explicit and newly visible, is what makes the pre-existing conflict known to all. Instead of focusing on a heroic narrative, diva citizenship redirects our attention to the importance of the spectacle of the conflict, the hypocrisy it will expose, and the publicity it will garner. The tensions between a public discourse of equality, democracy or even mere compliance and the reality of repression and resistance produce a new political space in which the question can again be raised as to whether the status quo is truly acceptable.

One of the most important points Rankine and Cohen make in their respective works is that the everyday-ness of the exclusion drives the nature of their political engagement. 'Most of these young black people not only see a black and white America; they would contend that they live and negotiate the consequences of such a racial ordering every day' (Cohen 2010: 120). Everyday life for many other Black youth is the deteriorating housing or being forced out of their neighbourhoods by gentrification, witnessing the violent deaths of friends and neighbours, high rates of incarceration of friends and neighbours, and one where they 'go to subpar schools, underfunded by the city, state, and federal government, few of whose graduates attend college. They live in neighborhoods where gang activity and the wrong policing strategies threaten the safety of neighborhood residents and significantly reduce their life options and choices' (Cohen 2010: 195, 231; see also Brunson 2007). Cohen argues this persists across class lines because of 'the systematic pathologizing of all black youth.…Surveillance and demonization of black youth are pervasive in this country and extend across class lines and across racial and ethnic communities, although clearly concentrated in poor communities' (Cohen 2010: 141). Rankine (2015) also makes it clear that there is no escape for anyone, as there is no corner of society or the media that is not infected with racist language and assumptions, undermining any idea of equal citizenship.

Embodiment and its vulnerability raises a further issue: the physical and psychological limitations of human beings. The daily injuries drain us of energy, time, focus, patience, all of which is needed to navigate everyday life, and this in turn increases the difficulty of our urban negotiations. They lead sometimes to disengagement, including from politics (Cohen 2010). Berardi underscores how we have pretended to ignore the limits of the body and its vulnerability to 'exhaustion': its 'limits of attention, of psychic energy, of sensibility' (Berardi 2012: 77). Social exclusion and trauma have robbed people of their well-being, joy, health, relationships, employment, and housing. The trauma is the result of exposure to structural and physical violence, to personal attacks, that are in turn the result of misogyny, patriarchy, militarism, imperialism – systems that hierarchically empower and oppress.

Just as acts of exclusion and violence, even when individual and intimate, are political, so too is the consequent need for acts of care, acts of compassion, acts of healing – of oneself and others. There are 'other relationships of ownership and belonging that exist' even if they are not formal or given recognition or protection by the State (Cooper 2014: 175). You can belong to a place – you can be constitutive of it and it of you – without owning it or having formal or authorized access to it. Disengagement, withdrawal, and refusal are logical and reasonable responses to the exclusion of the city and nation (Wood and Wortley 2010), and these moves do not lead into an empty space, but rather into more private, intimate and anonymous spaces where joy, pleasure and healing can be found. They often lead into community: 'As in the past, it will probably be the organizers, advocates, and communities outside the traditional power structure who will do the hard work of changing the life trajectory of marginalized black youth' (Cohen 2010: 232). They may be publically accessible third places, as in the examples of Toronto bathhouses (Cooper 2014; Nash and Bain 2007) or African-American barbershops and salons (Wood and Brunson 2011). These spaces and practices are everyday utopias, not only for safety, freedom from stigma and oppression, but for community-building, for love and desire and pleasure.

This imagining and creating of such spaces is also work that generates optimism and hope, which are also necessary for further new ideas. Resonating somewhat with Castells, Cooper argues that optimism and hope are needed because of the inherent, apparent impossibility of change. In the face of ideologies that trumpet the commonsense of the status quo and the idea that there is no alternative, the difficulty of asserting otherwise is not trivial. But the uncertainty of success may also be embraced for the potential that also resides in instability and through a refusal to accept that anything is predetermined: 'the future is no longer solidly and predictably connected to the present but erupts suddenly and without notice' (Cooper 2014: 219–220).

Art, play and the city: Acts of citizenship and healing

Finally, I want to consider these political understandings and acts in the context of the nature and experience of urban space. Acts of citizenship by invisible citizens are very much of, and suited to, urban space; they emphasize and extend its dynamism, flexibility, and ephemerality. Interruptions of the city's space and rhythms that stand outside capitalist market exchanges and challenge in some way the regulated, oppressing order of the city are commonplace. Revisiting a spectrum of these interruptions through the lens of diva citizenship, refusal, resistance and (re)creation demonstrates how omnipresent such practices already are in city space, and their consequent potential. Central to these acts are not just emotions like courage or hope, but the concepts of play and art.

Play is an important part of being serious. It is fundamental to the acquisition of new knowledge and problem-solving. Not unlike art, play enables the imagination to transcend the present and the real, even while engaging with the materiality and reality of the world as it is: 'play as an activity and attitude that involves creativity, some quality of pleasure, concentration, and absorption as it pulls participants out of their ordinary worlds into one with different rules, norms, moves, and concerns' (Cooper 2014: 200; see also

Huizinga 1970). Play also re-energizes ongoing political action, such as 'how DIY activist communities can reinvent otherwise tired, worn, and static modes of political and civic engagement' (Reilly 2014: 127). Reilly's study of The Yes Men, activists who stage media pranks and hoaxes impersonating corporate and government representatives, described how the group was able to generate 'actions that are at once funny, absurd, ethical and moral in scope – not to mention driven by the need to advance social justice issues…' (Reilly 2014: 131). Graeber describes what happened during a road blockade in front of an oil refinery: 'the access road became a party: with music, clowns, footballs, local kids on bicycles, a chorus line of Victorian zombie stilt-dancers, yarn webs, chalk poems, periodic little spokescouncils…' (Graeber 2011: 108). Play, music, and theatre have a long history within activist public protest. Here, I want to argue for the inherent political quality of several, less obvious examples of play in the city.

There are minor examples of changes to street use, which have little political intent, but nevertheless reveal how dominant capitalist hierarchies of use are, and how easily disrupted they may be (see Lefebvre 1991 [1974]). Something as simple as a 'scramble' intersection, which stops all vehicular traffic and prioritizes pedestrian crossing in all directions, turns the streets into a public square where movement is disorder and ceases to follow the lines of the grid. Charity runs and competitive marathons close roads and also put the exclusive use of (and the power of) those streets to the purpose of exercise, athleticism, fundraising and awareness. The fundraising is not the only thing accomplished – a public is reached, the urban consciousness of daily life of anyone trying to navigate the usual routes by automobile is interrupted with disease awareness or the needs of children, through activities that resemble play more than work.

Then there are similar acts of play, done with greater collective intent, such as flash mobs. These are often held in city streets or large third-place public spaces, such as malls, for maximum visibility. They are commonly recorded and posted on the Internet to extend the audience past the moment of performance. Flash mobs have an intense sense of play and draw some of their energy from the surprise of bringing the joy and energy of music and dance, or costume, or other playful, unusual behaviour on the part of a group into an otherwise mostly functional space of individual movement. It is a spectacle by and for ordinary people, turning the daily routine (social and economic, domestic chores, work routines) into a moment of joy for its own sake. The vulnerability inherent in the willingness to perform, even look silly, is disarming, and makes the city less formal, more welcome, more socially accessible. It also temporarily creates connections and even collectivity in the audience it gathers, and in the interaction between that audience and the performers.

Flash mobs are of course not the first use of the street for performance. City streets have been the stage for breakdancing, singing on streetcorners (with the use of streetlight as a spotlight), and a myriad of busker acts: magic, dancing, acrobatics. Spontaneous and regular use of certain spaces turns public space into places of temporary but routine occupation and performance – a territorial claim. As McCormack writes, 'the quality of moving bodies contributes to the qualities of the spaces in which these bodies move. Put another way, spaces *are* – at least in part – as moving bodies do' (McCormack 2008: 1823). These particular acts serve to entertain their audience but also to develop one's talents, improving and helping oneself in direct interaction with others, including (perhaps especially) strangers. Still more fundamentally, they often engage the body in expressive movement. Forms of dance or acrobatics generate multiple sensations for participant, such as physical, kinaesthetic, and affective, and the movement may be emotive, energizing, cathartic (McCormack 2008). Nor is this limited to 'ideal' bodies; work in critical disability studies has confronted the normalization of 'ideal' bodies through an emphasis on the social production of ability, and it has also argued for the importance of re-engaging the disabled body. Seeing artistic movement not merely as 'therapy' or 'rehab', but as 'expressive art', has been key to a reclamation of the disabled body 'that runs counter to the social expectation that divergent physicalities should be concealed from view' (Snyder and Mitchell 2001: 382, 385; Adewunmi 2013).

Streets and sidewalks are also the site of play of a different kind of performance. From their first encounters with them, children have been turning the city's hard surfaces into playgrounds for hopscotch, football, and other sidewalk/street games, and canvases for art. In these acts, they become in some ways stewards and managers, even governors of those spaces. They manage space and activity, taking turns, figuring out who participates, organizing activities, creating and resolving conflicts (see also Cooper 2014: chapter 7). They beautify it with chalk drawings. Through their own acts and often their own initiative, children govern their small part of the city. Through no formal or authorized process, these acts, especially those that leave a trace, secure a space for themselves, developing a kind of user rights, even as it may disappear each night or work day.

Another act that transforms city space is only known for the trace it leaves behind: the invisible artists who territorialize (tag), beautify and politicize city spaces through graffiti art. Their work relies on (and is thus a product of) urban rhythms, surfaces and spaces, the abandonment of sites, the 'privacy' of night, or other ways in which places and buildings are out of public view (McCormick 2011). This urban and suburban landscape is rich in smooth services that take paint well: transit vehicles and rail cars in storage yards, empty buildings, no-longer-functioning factories, the outside walls of office buildings, closed shops, empty parking lots, construction or demolition hoardings. The infrastructure of the city that is well suited to graffiti and muralism is not coincidentally in poorer and racialized urban neighbourhoods (Bofkin 2014; Deitsch 2011). It is not a coincidence that marginalized communities wound up next to railyards, public transit stations and tracks, or vacant lots, or that more and more of the urban fabric was left to decay as the mostly white middle-class left inner cities for suburbs. Urban renewal public housing projects in Europe and North America in the 1960s were often monolithic towers. Bofkin connects this urban social and economic 'destabilization' directly to the emergence of street art. Drawing on his observations of examples from North America, South America, southern Africa, North Africa, and Europe, Bofkin finds it almost universally demarcates a less well-to-do neighbourhood: 'Street art is culturally atypical in up-market areas for most cities around the world' (Bofkin 2014: 288).

Bofkin also draws a straight line between the earliest examples of street art, which were writing (usually signatures) and the urban landscape with its pervasive commercial signage. 'Given that the written language had become an everyday and expected part of the urban landscape – and nowhere more so than in New York – we can understand why someone would write their own name on a wall, or the side of a train' (Bofkin 2014: 12). Graffiti does not merely cover urban surfaces; it also interacts with them, using three-dimensional and anamorphic representations and other techniques of trompe l'oeil to distort the flat surfaces. Artists intentionally make physical features of buildings stand out or disappear, or use the city's physical form as part of the shape of the creation. 'The beauty of street art is that it teaches you to look at spaces not for what they are but for what they could be' (Bofkin 2014: 6).

Thanks to urban mobility and public transit, graffiti and street art travelled around the city, on the inside and outside of subway trains. The resulting fame existed within a network and social world that was invisible to much of the rest of the city. As Bofkin notes of Philadelphia's original hand-drawn sticker street art, 'it is invisible in plain sight, right under the noses of – but ignored by – most Philly residents' (Bofkin 2014: 172). The more bold, difficult or unusual the location, the greater the respect for the artist: 'trains are the most challenging surfaces to paint;…the works embody the spirit in which they were made: the audacity and the risk' (Bofkin 2014: 256; Deitch 2011). When the media paid attention, writers and artists selected places more likely to be seen and photographed, and oriented their work, even signatures, towards art. Importantly, the chosen location is a commentary in and of itself: of neglect, decline and crisis, even death, in the city. It makes that loss visible, draws attention to it, and constitutes a form mourning, as well as rejuvenation and rebirth as art.

Street art has often provided explicit political commentary. It may mark partisan space within the city, as well as express international solidarity with other struggles, such as the murals in Belfast (Rolston 2004; Dowler 2001); a new surge of muralism made the grief, anger and pride of the African-American community larger-than-life in Baltimore following the death of Freddie Gray in police custody (Giordano 2016); it served as an anonymous yet public canvas for angry satire against political and economic elites, as in Ireland in the wake of the crash (Meegan and Teeling 2010). The art is often elaborate, but key words, phrases and symbols commonly circulate as well. Once 'Occupy' had become part of public discourse, graffiti of the word alone became a political declaration on the surface of any building worldwide.

Consider all of the above as not distinct from, but on a continuum with, more visible collective public action, such as marches, rallies and occupations. Graeber's approach to the latter helps link the two, as he recognizes how practical, creative and playful aspects often merge (Graeber 2011). Castells has noted that even with all the magic of the Internet, there remains a need 'to build public space by creating free communities in the urban space' in part because it 'makes itself visible in the places of social life. This is why they occupy urban space and symbolic buildings' (Castells 2012: 10). He further notes the long history of such practices, their capacity to build community, their symbolic power, and the public creation of a space for deliberation, which is inherently political. But what is missing from most discussions of activism is the existence of these spaces in less public and explicitly collective ways. The spaces of autonomy that live in private and third places, the ways in which the city is reclaimed more privately or anonymously – collectively, yet not as a crowd or 'the people'.

These acts are all ephemeral (except perhaps graffiti), and street artists often embrace this quality rather than resist it, such as with flash mobs or those who use fragile materials, such as fabrics, yarns, paper and tape which will not last long against the weather or urban traffic. Some artists specifically prefer the embodied immateriality of performance and reject even recording, arguing that, like so many other aspects of urban life, '[t]hese artistic acts are not meant to last beyond their execution' (Bofkin 2014: 174; see also Grubbs 2014). These works celebrate improvisation and experimentation, and destabilize ideas of authoritative performances or permanence – of the art, and of the places it is situated whose order is disturbed.

Assessing the value or impact based on the time scale of an event or act is limited and unproductive. This work is 'temporary', as all movement within the city is. What do we mean by temporary, and what political value are we embedding in the idea? After all, the city's functioning is largely movement, a series of temporary yet durable patterns. These acts are also enabled by the possibility of the random that is the built and lived structure of the city. The city relies on randomness and makes space for it. So much of the city is a series of channels that are (necessarily) designed to allow for unscripted movement, for the diversity of paths that its residents and visitors carve and follow. While the city's design – through the acts and plans of police, politicians, property owners – also seeks to channel people in particular ways and restrict certain paths, the city also creates all kinds of opportunities in terms of time and space to express oneself, leave one's mark, to shout.

There is an argument for seeing graffiti 'as a reflexive beautification project' (McCormick 2011: 22). As a volunteer at a major street art festival, Meeting of Styles, put it:

> For me, graffiti is an expression of my desire to make the world a better place. It's a colorful revolution. The Meetings are my contribution to help

> build a better society, in which creativity rules and is considered more valuable than materialism and senseless consumption.
>
> (Bofkin 2014: 265)

From its moving bodies to its hard surfaces, the city is saturated with politics. These acts of play and works of art are all acts of citizenship. Their authors are scared, sad, excluded, angry, traumatized, tired, ill, lonely, outraged, grieving, happy, playful, desiring, lacking, and their interventions into the urban landscape deliberately disregard, transform, transgress or disrespect (even attempt to destroy) boundaries of usual or appropriate behaviour. In these acts, they call out its false order, its lifelessness, its injustice. They increase the fluidity, flexibility, and dynamism of the city, returning attention to the present moment, the present place, and its potential. Many of the authors or participants are anonymous: members of a crowd, unknown and unseen, or intentionally invisible, seeking to 'avoid objectification, regulation, and rationalisation' (Bain 2004).

And these are acts of citizenship for another reason: art, music, dance and other movement or exercise are all therapeutic. They are acts of care that help people to survive (Salek 2014; Hedva 2016; Lorde 1988) They are processes of healing, becoming stronger, becoming well and whole; they provide relief from psychological and physical pain in the reassertion of the self that is being and becoming. Berardi sees some of this as critical to 'the uprising':

> The uprising is a therapy....The uprising is not a form of judgment, but a form of healing. And this healing is made possible by a mantra that rises, stronger and stronger, as solidarity resurfaces in daily life.... [T]he task of the movement is to act as a medic, not as a judge.
>
> (Berardi 2012: 133)

These are expressive acts, they have strong potential for connection with others, and they are an effective means for the production of positive social space. Art and other creative work should be the source material for much more work on citizenship than it is. Here I am limiting myself to popular, informal, and illegal forms in urban space, but not because I think that is a complete list of the role of art in urban political protest and activism. Art, music and physical movement are political acts not (only) because of their power to disrupt, challenge, make political statements, etc., but also because of their therapeutic power to soothe, vent, beautify, empathize, support, encourage, give hope, and heal, for both the actors and their audiences.

The therapeutic, sensory impact affects the city as a whole. As Berardi reminds us, 'Sensibility is the ability of the human being to communicate what cannot be said with words' and he argues it is productive of 'social solidarity' (Berardi 2012: 121). What this sensibility – which includes art and play – achieves is a kind of decompression of time and space, as it engages the senses in a way that cannot be fully processed immediately, and in a way that lingers. 'Sensibility slows interpretation procedures, making decodification aleatory, ambiguous, and uncertain, and thus reducing the competitive efficiency of the semiotic agent' (Berardi 2012: 127). Art changes its environment and its artist. Turning a critical eye on how the world of global finance has disconnected language from material reality, Berardi believes that resistance starts with reclaiming language (he notes that symbolist poets did the same thing with language, cutting it off from its material referents, although in a different spirit). He looks in particular to poetry, for 'the intimate ambiguity of the emotional side of language' (Berardi 2012: 21), language that does not concern exchange, and goes beyond functionalism. Poetry's refusal of efficiency of meaning, clarity, linearity, and one-to-one relationships stands in opposition to language as information. The poet Dominique Christina notes 'how delicious and possible language is' (Acevedo 2016). As with other forms of art, poetry's intentional multiplicity and complexity of meaning compels the reader to slow down. Through this regeneration of sensuousness and richness of language, Berardi believes we can find

> a new common ground of understanding, of shared meaning: the creation of a new world….This place we don't know is the place we are looking for, in a social environment that has been impoverished by social precariousness, in a landscape that has been deserted….It is the place of occupation, where movements are gathering: Tahrir square in Cairo, Plaza do Sol in Madrid, and Zuccotti Park in New York City.
>
> (Berardi 2012: 147, 149)

The Baltimore poet/spoken-word artist, Kenneth Morrison Wernsdorfer, is an example of the same kind of thinking, although it is not from neoliberalism's financialization of language that he needs to reclaim a vocabulary.

> I can read [Robert Frost's] work, and feel better, or think about life from a place of hope. Whereas as an artist, I have to write from places of deeper trauma. But I also assume Robert Frost and I live – lived – in completely different worlds and different realities, where he can write about, you know how beautiful a rose may be. I have to write about police brutality.
>
> (Randolph and Bowker 2016)

Wernsdorfer runs a youth programme of creative writing, *Dew More Baltimore*, that plays a critical role in, among other things, helping young people process trauma. Dominique Christina's politics and art also connect directly to trauma: 'I bleed out loud' (Acevedo 2016). For Wernsdorfer, 'the role I think art plays in this is raising awareness, challenging people's comfort, giving people new tools, new frameworks, new ideas of how to look at this kind of work'. For Christina, 'Writing…is about the insistence of myself, the naming of myself, and the defense of myself' (Christina 2015: 5). Both poets take an intersectionalist approach to racism, the complex and tight intersection of several institutions that structure marginalization and oppression. An equally complex response is necessary, and no one act or challenge will undo it all. To dismantle power 'requires a reframing of things, and certainly the decolonization of our minds' (Christina, in Emmanuele 2015). Wernsdorfer believes the creation of art is the creation of imaginative places where vulnerability can be acknowledged and protected. Art is a necessary and political act of urban citizenship, where justice requires finding ways for the suffering to speak.

> For me, saying anything at all is an act of resistance. I'm not supposed to be here. I'm not supposed to know what I know....I am deliberate, therefore, about standing in the center; naming and claiming and telling truth to power.
>
> (Christina, in Emmanuele 2015)

Conclusion

The sad, sick or diva citizen is a suffering citizen, but also a creative, productive subject whose political interventions are agonistic (in every sense), and thus insightful and innovative. This citizen is fundamentally and clearly a producer and product of the surrounding society and landscape, not an abstract (or abstractable) individual. This constituted political subject who inhabits an actual city should expand our understanding of politics – refusing a politics of only formal institutions, of civil debates, of processes of governing and resistance (both of which suffuse our daily lives and spaces), of the individual and unaffected body. We have only individuated bodies within the social body. The city's interdependence makes this reality stronger and clearer – it is fundamental to its economy and society that people move and exchange, that the survival of the individuated person or family is reliant on the collective.

Experiences of, and responses to, the citizenship gap are complex and multiple. Making the citizenship gap visible and known in the first place may be its own accomplishment. 'The populations who were and are managed by the discipline of the promise – women, African Americans, Native Americans, immigrants, homosexuals – have long experienced simultaneously the wish to be full citizens and the violence of their partial citizenship' (Berlant 1997: 19). The citizenship gap is glossed over, and what falls into the gap is what we might call the collateral damage of our political posturing. Its consequences are material, even a matter of life and death. When Cohen writes of the failure to provide Black youth with the environment, resources and opportunities they deserve, she asks, 'Are their lives not worthy of protection? Are they not full members of our political community? Does no one care about them?' (Cohen 2010: 15). Her language of 'protection' and 'care' evokes citizens and the nation as emotional bodies, and frames questions of justice and fairness as concerning suffering and wellness.

Our studies of social movements, of protest, activism and resistance are heavily weighted by what we see in public and how it is framed as 'legitimate' or 'illegitimate'. We focus on the banners and slogans and mass protests and organizing meetings/public assemblies, but not the motivations. We then theorize their politics at the scale of the assembly; even if we know and recognize different groups are present for different reasons, we focus on the solidarity, what they hold in common that leads to the alliance. We do not ask who is not present and why. But the social geography of the inhabited city tells us. We cannot have expectations of solidarity until we recognize what lies behind the fractures and fragmentation of the city. Those who wish to be allies, such as white feminists who normalize their experience as representative of all women, would do better to dig more deeply into their own suffering, consider their commonalities with and the greater complexity and suffering of women who cannot exercise white, able-bodied, cis, housed, settled privileges, rather than exploiting the advantages of their privilege to jockey for a relatively better position (Crenshaw 1989).

If we heed the issues raised by sad, sick, diva citizens, our means of measuring or assessing closure of the citizenship gap should include the relief of suffering: the cessation of violence, the avoidance of surveillance, the end of deprivation, the end of exploitation, secure and safe housing, the improvement of neighbourhood conditions, safe and secure access to needed resources, respect and dignity in public and private, relationships that are freely chosen, equal opportunity for joy. This is a geographical question, for the means to achieving the relief of suffering includes access to space for social and political gathering and organizing, for play and improvisation, and for joy and love. We cannot create those positive spaces without understanding and incorporating into our analyses the barriers to such spaces, as well as the way they already exist throughout the city.

4 The arc of politics

How do we come to terms with early twenty-first-century urban activism and protest in all its diversity? How do we build on acts of diva citizenship and healing to shape critical urban theory? In this final essay, I review the strengths and limitations of Marxian political economy approaches to critical urban theory and compare them to the possibilities that an anarchist approach presents. I treat Marxism and anarchism as two types or broad categories of politics, which differ not only in specific strategies and targets, but also more fundamentally (and more importantly for me here) in their understandings of the time, place and scale of politics (see also White and Williams 2014, who emphasize anarchism as a form of organization). The anarchist view has a less linear, less compelling and less insistent sense of time. Marxian analysis of the city has focused on the political economy and its shaping of the city; its desire for justice in that framework zeroes in on the dismantling of the capitalist system. With a stronger emphasis on praxis, anarchists inhabit, theoretically and literally, the nooks and crannies of the city, embracing its fluidity and capacity for change, and trying to transform the city by expanding those spaces where social relationships are not mediated by capital or oppressive hierarchies. For Marxian theorists, history and time have a weight, and theory has a consequential completeness. The nature of anarchist theory and praxis produce not only a different pace and strategy for change, but a fundamentally different understanding of political time and space.

I will not rehearse points of dispute between Marxists and anarchists further than this. I realize there are many ways in which they disagree and challenge each other, and that differences of opinion on critical points exist among Marxists and among anarchists, as well (for more on this, see Springer 2014 and responses, as well as Harvey 2017 and responses). As I compare Marxism with anarchist thought, especially that of Emma Goldman, I do not mean to present Marxist scholarship as doctrinaire and I think the broader term, 'Marxian', to describe the bird's-eye-view, systematic analysis of urban political economy, is more appropriate to this discussion. Clearly there are Marxist scholars who have brought out different forms of injustice (e.g. capitalism and patriarchy) as distinct though mutually enabling. Nevertheless, similar to the limitations within the debate on the post-socialist moment or the post-political condition, the Marxian approach continues to lack an engagement with diversity that might provide a different starting place. There is an extensive body of experience and theory regarding Marxists' failure to embrace feminism and other radical politics along with emancipation of the working class (see, for example, Rose 1993; Deutsche 1996; Katz 2006; Jarrett 2015). Here, I want to emphasize the value approaching the city as it is inhabited, with a recognition of its embodied diversity and the difference that makes. Its wider range of modes of resistance and a grounding of politics in the relief of pain and suffering lead to different understandings of 'revolution', different experiences and definitions of governance, and thus fundamentally to different definitions of what it means to be a citizen.

Scholars continue to parse and rank political acts. Katherine Rankin, for example, has distinguished between 'subversion' and 'resistance', arguing that resistance is 'collective, overt actions' with intent, and subversion 'more ambiguous political agency – individual, covert instances of nonconformity that engage tactics to get as much as possible out of a constraining situation' (Rankin 2011: 110). Some scholars have tried to include these less purposeful, even invisible, acts through ideas like 'primitive resistance', 'infrapolitics' (Scott 1985, 1990; Moreiras 2004) or adding other prefixes to 'politics' (Rancière 1999; Žižek 1999; Bosteels 2010). Scott's specifies infrapolitics as 'resistance that avoids any open declaration of its intentions' (Scott 1990: 220). Vodovnik (2013) uses infrapolitics as part of his own analysis of, and as a descriptor for, anarchism. These nuances and modifiers offer insight into different forms of politics or political acts, but I am not going to use them here to describe acts that are invisible or inscrutable to some, because in this context it would privilege certain forms of resistance over others (see also Leitner et al. 2008 on privileging scales of politics). Moreover, these specifying prefixes do not address the challenge of intersectionality too often missing from our understandings of being political. I prefer not to modify nor place differentiated value on different acts along the spectrum of resistance. I think it is problematic to guess at the value or intent as we would need to – similar to Butler's concern as to which and whose politics were deemed 'merely cultural' and why (Butler 1998). I believe not only are all these acts politics, they are a very important form of politics that could be considered revolutionary (if we want to continue with that word) and a form of constitutionalism, as it can be argued they are an effort to 'alter the cultural and legal self-understanding of their political communities' (Kalyvas 2008: 4).

All the communities and constituencies discussed in earlier chapters, whose politics are grounded in particular experiences of inhabiting the city, seek transformation and change. Measuring the value of their efforts, their goals, and what constitutes politics and resistance remains a problematic discussion among activists and scholars (to which the mainstream media also contributes). 'Only by thinking of citizens as being governed is one able to understand resistance in a constitutional way, or perhaps as something that resembles the *right to resistance*' (Duso 2010: 86). I propose that anarchist theory from its earliest conceptions had productive insight into governance and resistance. Not coincidentally, it was literally developed on the streets. I want to build on Emma Goldman's work and offer anarchism as a broad theory – not a master theory – of radical political practice, within which specific theories and practices may be developed and revised for specific communities and contexts, in the nature of anarchist praxis. The intention is not to identify any or all activists as anarchist, nor to suggest that they actively or otherwise seek the removal or devolvement of the State. Rather, it is to argue for an anarchist type of politics as a way of understanding these citizenship practices in the city, and their relationships to previous movements. This frame offers productive ways of measuring the space of politics, the space of the city, and the breadth of discourse, and brings a richer and more concrete understanding of new political strategies in the city, and how, in turn, these new approaches to post-crisis activism and protest bring new insights into the literature in critical urban theory and radical politics about what it means to be political.

The politics of critical urban theory

Critical urban theory is entangled in the same questions and struggles as the Left as a whole. Identifying a need for new approaches, a 2011 collection of essays by leading scholars attempted to address cities as transformed by global, neoliberal capitalism, and determine a path to more just and liveable places. The authors engaged a variety of frameworks, and the subsequent essays illustrate well both the insights and limitations of different approaches to critical urban theory. As Brenner, Madden and Wachsmuth declared:

> It is, we would argue, certainly not a moment for intellectual modesty or a retreat from grand metanarratives, as advocated by some postructuralists a few decades ago. On the contrary, from our point of view, there is today a need for ambitious, wide-ranging engagements – theoretical, concrete, *and* practical – with the planetary dimensions of contemporary urbanization across diverse places, territories, and scales. Yet, it would be highly problematic to suggest that any single theory, paradigm, or metanarrative could, in itself, completely illuminate the processes in question.
>
> (Brenner, Madden and Wachsmuth 2011: 118–119)

These scholars are aware of the depth, complexity and intransigence of urban inequality, and seek to produce 'a critical urban theory that is capable of grasping our global urban world "by the root"' (Marx 1963: 52; Brenner, Madden and Wachsmuth 2011: 134). As Mayer argues, too many political projects, such as the *World Charter for the Human Right to the City*, are limited in their thinking: they 'boil down to claims for inclusion in the current city as it exists. They do not aim at transforming the existing city – and in that process ourselves' (Mayer 2011: 74).

According to Brenner, 'Critical theory is...not intended to serve as a formula for any particular course of social change'; it should open our eyes to the problems and the possibilities, through both abstract thought and what we learn from practice (Brenner 2011: 15). For guidance, he returns to the Frankfurt School. While he recognizes that 'this "urbanistic" reorientation of critical theory will require further theoretical reflection, extensive concrete, and comparative research...' (21), Brenner's critical urban theory considers only how global capitalism and neoliberalism have shaped the city, and what he presents as grounded experience and perspective of the city is limited. He mentions Fraser's feminist critique of critical theory, but does not integrate it into his own argument.

While capital's power in the city must not be overlooked, it is also urgent to recognize and incorporate into an analysis of the city the power of whites, men, the enabled or 'able-bodied', cis-hetero persons, settlers and so on. Cities are built explicitly to preserve forms of social power (Deutsche 1996; Goonewardena 2011). Some of the scholars in the collection explicitly acknowledge other forms of oppression. Marcuse, for example, recognizes those 'persecuted on gender, religious, racial grounds' (Marcuse 2011: 30), as well as a general 'right to produce the city' as well as 'consume it' (36). Schmid notes that the city is 'a place where differences encounter, acknowledge, and explore one another, and affirm or cancel out one another' (Schmid 2011: 49). He recognizes how social movements have both resisted and reinvented the city, that they 'have created many kinds of concrete urban spaces and alternative, oppositional everyday practices, often based on cultural, ethnic, or sexual differences' (53). Rankin's approach to the city incorporates 'the logic of capital accumulation and *other modes of domination*', including 'processes of imperialism, racialization, [and] male domination' (Rankin 2011: 103, emphasis added). These remain, nevertheless, rather muted ways to discuss phenomena such as racism, colonialism and violence. Citing Fraser's efforts to develop 'a theory of justice that engages a politics of recognition in conjunction with a politics of redistribution', Rankin hopes for marginalized peoples to 'come to recognize the arbitrary foundations of prevailing systems of exclusion as well as interests in common with those who are differently marginalized' (111). Echoing Fraser's fears, she believes that 'recognition' has displaced 'redistribution': 'most formulations of cultural diversity in planning theory...ignore the socioeconomic base of so much cultural difference' (112). Redistribution remains at the core of these analyses, and their incorporation of recognition is more abstract than concrete.

Yiftachel's focus is on less visible, socially marginalized communities, and he makes the critical observation that both their tactics and their goals often differ from less marginalized activists:

> The new politics often distance identities and mobilizations from the state, signaling the fragmentation of the apparatus of power 'from below.' They often begin with struggles for 'insurgent citizenship,' as identified by

> Holston (2008), but may go further and transform into struggles for multiple sovereignties....[T]here is a point in the struggles when citizenship, integration, and equality...are no longer the dominant goals, but are intertwined with efforts to create autonomous ethnic spaces of development and identity.
>
> (Yiftachel 2011: 152)

Yiftachel observes that marginalized communities' 'agonistic opposition is likely to shift to antagonistic radicalism, and the horizon of equal integration may be challenged by an agenda of autonomous disengagement from the societal mainstream' (Yiftachel 2011: 158). He argues that critical urban theory commonly overlooks this differentiation of identity and power, and instead assumes equality and citizenship as a fundamental premise. He notes the particular absence of colonialism.

To even start to address the politics of marginalized people in the city, scholars need to recognize the limitations of Marxist thinking on what constitutes social groups and their politics, to allow for a political field constituted by more than economic interests (Kalyvas 2008). This is not a new critique. Weber found Marx's approach to politics as weakened by his 'class essentialism', and proposed thinking about the motivations driving the formation of political organizations as a psychological process before seeing them as a desire for power. For Weber, 'Politics is a quest for dignity, an assertion of identity first and then only secondarily an attempt to conquer the state and to realize one's interests through political means' (Kalyvas 2008: 44). This provided a break from a purely structuralist and economic approach to political identities, and emphasized that there is a *process* at work and not an automatic equivalence of class (structure) with group (aware, and possibly mobilized). Weber argued that this process of constituting political identities 'involv[ed] contingent, conflictual, and contest-specific relations that unfold in the sphere of the symbolic' (Kalyvas 2008: 45). 'Material' should not be seen as prepolitical and deterministic. The material context of any given struggle shapes the production of identities, and vice versa.

One element of a critical urban theory that is diverse and inclusive is a groundedness in the concrete and in everyday life. Like Purcell (2013), Schmid returns to Lefebvre's belief that 'the point of departure of critical social theory should always be everyday life, the banal, the ordinary. Changing everyday life: this is the real revolution!' (Schmid 2011: 58). Goonewardena agrees that '[t]he city is thus always a concrete, practical experience, a place of its residents who use it and appropriate it in their everyday practices....As such, everyday life also ought to be the central concern of any radical urban theory' (Goonewardena 2011, 98; Goonewardena et al. 2008). Rankin also proposes to place everyday life and its challenges at the centre of planning theory, in such a way that it can 'expose the common causes of peoples' day-to-day struggles everywhere as a basis for building a constituency for the collaboration and solidarity that will be necessary to push for a better economic system in which the demands of all can be met' (Rankin 2011: 106–107). Even with an emphasis on redistribution, the importance of the concrete experience of everyday life and its revolutionary potential, as Lefebvre advocated, is increasingly prominent in critical urban studies.

Yet Yiftachel argues that, for the most part, critical urban theory has not gone far enough to 'account[] for the implications of a new political geography, characterized by the proliferation of 'gray spaces' of informalities…, forming pseudo-permanent margins of today's urban regions, which exist partially outside the gaze of state authorities and city plans' (Yiftachel 2011: 153). Yiftachel is open to how consideration of these geographies will transform the frame of politics through which critical urban theory can be developed. He recognizes that the dynamics of these communities, including their social invisibility and exclusion, compels 'the politics of identity as a central foundation of urban regimes, intertwined with, but far from subsumed under, class or civil engines of change typically highlighted by CUT [critical urban theory]' (152). Yiftachel's use of 'new' here frustratingly echoes the limited perspective of the post-political literature, but the engagement with identity is substantive. He notes that grey spaces are an increasingly common phenomenon in the developing world, but they are not unique to those regions. In the 'developed' world, we could also include, for example, what Joseph Heathcott termed the 'Black archipelago': 'islands of vibrant black social life surrounded by seas of white racism and hostility, cities within cities that stretched in a chain across America' (Heathcott 2005: 707); the encampments of Travellers or Roma; queer third places; and unauthorized 'tent cities' (which is closer to what Yiftachel has in mind).

There are many strengths in this literature, one of which is an explicit desire to diversify the idea of who is a political subject. However, the form of future activism is assumed to be a coalition or convergence, some kind of unity, and often leaves aside the difference that the diversification of who acts as a citizen will make. This leaving aside creates interesting tensions: for example, in Rankin's essay, between feminist advocacy for 'change begins "at home" with everyday practice and experience' and yet advocacy for moving to a larger field that, in practice, negates a politics rooted in everyday life; between 'we understand difference in terms of historical relationality and accountability rather than as static, embodied categories' and a politics that seeks to transcend that specificity, 'building solidarities across difference out of which can emerge *stronger* theorizations of *universal* concerns' (Rankin 2011: 113, emphasis added). Marcuse acknowledges that 'the view that it will be the proletariat that, as a single class, leads the struggle…is outdated' (Marcuse 2011: 34). But 'the struggle' is still singular, and he hopes to see 'a convergence of all groups, coalitions, alliances, movements, assemblies around a common set of objectives, which see capitalism as the common enemy and the right to the city as their common cause' (34). Schmid argues, 'The various alliances that have coalesced around the rallying cry of the right to the city demand…*a new unity* in the splintered and fragmented urban regions' (Schmid 2011: 58, emphasis added).

There is in these accounts insufficient engagement with intersectionality, particularly as it is embodied and lived, and too little consideration of the toll of hate and exclusion on targeted individuals and communities, as well as the difference such perspectives and experiences makes in theory and practice, including among 'allies' on the Left (We Are the Left 2016; Peake 2016b). Ideas of solidarity, while welcome, sometimes suffer from programmatic thinking about mass mobilization and coalitions. Derickson (2016) returns to Fraser to argue again for the need specifically in critical urban theory for engagement with the politics of recognition, precisely because the unified or 'totalizing' political subject serves to reproduce cultural misrecognition. A Marxian theoretical framing does not incorporate all the oppression and suffering of the inhabited city, and she underscores the need to find a theory and a politics that is capable of attending to the actual struggle of people's everyday lives. Moreover, this must go beyond mere inclusion: the inclusion of marginalized communities in scholarship developing critical urban theory often does not shift the perspective. More research with a focus on the details of everyday life of the inhabited city would be welcome, as well as a consideration of how intersectionality and its common 'jockeying for position' might frustrate the desired convergence (Crenshaw 1989). Instead there is a rapid move towards unity or convergence, which revolves around a narrow view of economic and political power. This is not a nuanced understanding of the reality of everyday life in the city, nor its politics. The city is not a unified place, but is productive of difference (Castells 1983; Deutsche 1996; Isin 2002).

Linda Peake (2016b) identifies similar limitations within theoretical framings of 'planetary urbanization' (Brenner and Schmid 2012; Brenner 2014). While it remains important to conceive of the urban as a process that encompasses more than cities themselves, we need to remember the process is always incomplete (Roy 2016) and incomplete within the city itself (Peake 2016a). As it purports to be 'the urban theory to encompass all critical theory', Peake argues that planetary urbanization is also incomplete, for the inhabited city remains absent from its theorizing (2016b: 831). We need 'praxis underpinned by a feminist mode of situated knowledge production' (833) in order to maintain our connection with the concrete, and allow those everyday struggles to inform what critical urban theory and radical politics looks like (see also Leitner et al. 2008). Mayer sounds a further cautionary note: 'Northern theories which imply that urban movements today must organize on a global scale – as, for example, David Harvey claims – are frequently rejected by movement representatives from the global South'; she suggests that something other than massive unity must be considered, citing De Sousa Santos's conclusion that the 'vast differences in both practice and theory… cannot be synthesized' (Mayer 2011: 79, 80).

Marxian thinking on revolution and time is similarly limiting for critical urban theory. Notwithstanding the contribution of Marx's analyses of the value of time, the sense of historical time is much less developed. 'In the first pages of *Capital*, Marx explains that value is time, the accumulation of time.

Time objectified, time that has become things, goods, and value' (Berardi 2012: 86). But his historical time turns on a belief in the revolutionary moment as something that changes the course of history, that there is a course of history that drives through the present at the future. As Berardi argues, 'Dialectical materialism, the philosophy of the past century, implied a form of moralism: anything (progress, socialism, etc.) that moves in the direction of history is *good*, whatever opposes the movement of history is *bad*' (165). Graeber also raises questions of a theory of history where there is a clear, linear (and implicitly inevitable) progress of one idea replacing another entirely. 'In part', he argues, 'this is the legacy of Marxism, which always tends to insist that since capitalism forms an all-encompassing totality that shapes our most basic assumptions about the nature of society, morality, politics, value, and almost everything else, we simply cannot conceive what a future society should be like' (Graeber 2011: 102).

The assumptions in Marxian thought – sometimes conscious, sometimes presumed – of the necessary scale of political and social change, its completeness and finality, are problematic. These assumptions come to the fore in the remarks of Harvey, Chomsky, Žižek and others in their evaluations of the Arab Spring, 15M and, especially, Occupy. They move very quickly to embrace and celebrate that 'kind' of politics, while they ignore others (both big public protests/social movements like Idle No More, Black Lives Matter, Slut Walks, etc., as well as the more invisible, amorphous acts discussed in Chapter 3), and they start to make a Plan. The desire to scale up is urgent and immediate, and entails the (problematic) construction of a collective, solidaristic identity that unifies the movement. The other weakness here is a misrecognition of Occupy itself. As several of those who have considered the occupations more carefully have noted (Graeber 2013; Williams 2012; Gerbaudo 2012; Castells 2012; Sharp and Panetta 2016), these events had more in common with anarchist political practices than with organized labour, and they were products of their particular cities. Ironically, this push for unity is uncomfortably similar to the invented patriotism of the nation-state. We have the usual 'good citizen', but this time, it is activism's good citizen. It encourages putting aside other identities, other needs and issues, for the 'greater' collective good. But the premises of that collective good are not questioned as they need to be.

In practice, there is a ranking of issues and needs. In some ways, these are old complaints. Feminists in socialist movements are familiar with the sense that they are to wait their turn, and it never seems to come. On the Left, feminist theory has similarly been marginalized and has made a limited appearance in the work of most male theorists (Graeber 2011; Peake 2016b). Within feminist theory, White feminism has too often (implicitly and explicitly) put racism and intersectionality on the 'back burner', neglecting or dismissing the issues and forms of political activism of Black women, women of colour, disabled women, lesbians, queer women and the trans community. A radical politics that incorporates a diversity of issues and allows for self-separation of groups as they see fit will need broader understandings of what constitutes politics, what activism may look like, and how transformative possibilities are imagined in time and space.

Anarchist theory and the politics of the inhabited city

In *Weapons of the Weak*, James C. Scott directly challenged narrow and normative definitions of resistance:

> *Real* resistance, it is argued, is (a) organized, systematic, and cooperative, (b) principled or selfless, (c) has revolutionary consequences, and/or (d) embodies ideas or intentions that negate the basis of domination itself. Token, incidental, or epiphenomenal activities, by contrast, are (a) unorganized, unsystematic, and individual, (b) opportunistic and self-indulgent, (c) have no revolutionary consequences, and/or (d) imply, in their intention or meaning, an accommodation with the system of domination.... This position, in my view, fundamentally misconstrues the very basis of the economic and political struggle conducted daily by subordinates – not only slaves, but peasants and workers as well – in repressive settings. It is based on an ironic combination of both Leninist and bourgeois assumptions of what constitutes political action.
>
> (Scott 1985: 292)[1]

If critical urban theory is to overcome such false and privileging distinctions and include the political struggle of everyone, it needs to move beyond Marxian and liberal ways of thinking. I propose that the work of the anarchist Emma Goldman may be productively used as a path to a more radical politics for the city – that is, the city as it is actually inhabited by a diversity of people whose relationships to each other are commonly antagonistic, and whose means of resistance are often unseen or dismissed, even by fellow activists. Goldman's thinking gives us a model for a comprehensive frame that includes different modes of oppression without prioritizing them, and identifies a means of solidaristic connection among citizens as emotional, embodied subjects.

Goldman defined anarchism as 'The philosophy of a new social order based on liberty unrestricted by man-made law; the theory that all forms of government rest on violence, and are therefore wrong and harmful, as well as unnecessary' (Goldman 1969 [1910]: 56). Goldman's ideas of 'man-made law' and 'forms of government' extended past the State. She centred her arguments around economic justice; however, she explicitly contextualized economics in a broad sense, and understood injustice and oppression as arising from multiple sources, including the State, religion, property, society, and the power of the wealthy. 'Anarchism therefore stands for direct action, the open defiance of, and resistance to, all laws and restriction, economic, social, and moral' (71). Consequently, Goldman's definition of emancipation included not only freedom from oppression and want, but the freedom to experience love, joy and fulfilment. 'Anarchism stands for a social order based on the free grouping of individuals for the purpose of producing real social wealth', which she argued 'consists in things of utility and beauty, in things that help to create strong, beautiful bodies and surroundings inspiring to live in' (68, 61). Goldman's anarchism fought any and all forces that prevented living the best possible life: 'the spirit of revolt, in whatever form, against everything that hinders human growth' (69).

Goldman was emphatic about the beauty of love, 'the strongest and deepest element in all life' (Goldman 1969: 242), but condemned marriage with equal strength, equating it to 'that other paternal arrangement – capitalism' (241), and to prostitution: 'it is merely a question of degree whether she sells herself to one man, in or out of marriage, or to many men' (185). The harms of both marriage and prostitution, Goldman argued, are the result of the unequal economic treatment of women, the religious (often specifically Puritanical) moralizing that keeps women in ignorance of sex and sexual pleasure, and the reduction of women's value to a 'sex commodity' (190). For women, the home is but a 'modern prison with golden bars' (202). Unlike many of her male anarchist contemporaries, she advocated for access to birth control and abortion, as well as basic sex education. Goldman acknowledged the specificity of women's issues, and the role of legal and social, material and discursive forms of oppression. As her biographer Ellen Ferguson wrote:

> Goldman's feminism was very much focused on bodies….Her integration of feminism into anarchism produced early versions of some of the central concepts of second- and third-wave feminism: opposition to dualist thinking, rejection of hierarchy in favor of intersectionality, and an early encounter with the equality/difference debates. She discusses what Foucault and others call the productive, as opposed to merely repressive, aspects of power, the specific functions of biopower and the intimate as well as structural consequences of what later feminists call the protection racket.
>
> (Ferguson 2011: 250)

Goldman refused to prioritize suffrage – not because women are any less deserving of participating in political decisions than men, but because Goldman believed the State is responsible for oppression and voting only perpetuates its legitimacy. She saw no evidence of the introduction of women's suffrage improving the State's support for women or other oppressed communities and individuals. She also repudiated the common arguments of the time of women 'purifying' politics as more of the moralism that oppresses women, and criticized suffragists for using such arguments to degrade and exclude immigrants, prostitutes and other working-class women. Furthermore, given the multiplicity of the sources of women's oppression, she argued the State cannot remediate many forms of their suffering – nor anyone's: 'all existing systems of political power are absurd, and are completely inadequate to meet the pressing issues of life' (Goldman 1969: 205).

For Goldman, the different forms of oppression called for different responses and strategies, including the responsibility of the oppressed individual to assert herself. Her thinking also recognized intersectionality, and she called out middle-class women for seeking to empower themselves through the normalization of bourgeois attitudes that increased the oppression of less privileged women. Her politics was praxis, and a direct product of her own experiences and engagements with people in their own struggles. Ferguson has called Goldman's politics 'ectopic': 'out of place, located in subaltern places, not the expected spaces of the state or the academy' (Ferguson 2011: 278). Ferguson situates her

> not within the dualism of philosopher vs. activist, but within a different register of political thinking....[I]t is situated, event-based, and concrete. She was most insightful when she was thinking in the streets.... ...Goldman's political thinking encourages us to imagine, not only universities and the library of advisors to those who rule, but union halls, kitchens, neighborhoods, unemployment lines, prisons, brothels, and other less conventional sites as spaces of political thinking.
>
> (Ferguson 2011: 6)

As a direct consequence of the diversity of spaces in which she worked and their specific contexts, Goldman presented no master theory of oppression or form of resistance. She recognized power and oppression in many forms and circumstances, and held that no form is acceptable, that all relationships should be non-coercive. Achieving such relationships and resisting coercion does not follow a prescribed path:

> Anarchism is not...a theory of the future to be realized through divine inspiration. It is a living force in the affairs of our life, constantly creating new conditions. The methods of Anarchism therefore do not comprise an iron-clad program to be carried out under all circumstances.
>
> (Goldman 1969: 69)

Difference was to be accepted and hierarchies resisted, wherever they were manifest. Moreover, as Ferguson summarized, there is no 'vertical structure in which an alleged primary cause underwrites all others. Since there is no ultimate cause of oppression or foundation of order, there is no final struggle, either, but rather an interlinked network of struggles against "all forms of tyranny and exploitation"' (Ferguson 2011: 261). To describe anarchism's many targets, Vodovnik uses a metaphor of fighting 'the mythological many-headed monster Hydra, wherein the state represents only one of the threats (i.e. heads) to be "severed"...' (Vodovnik 2013: 10; see also Shannon et al. 2012).

There are countless ways to control, exploit, humiliate and degrade, and an equal number of ways to resist. For Goldman, the work of resistance must start with a recognition of and sympathy for suffering:

> one must feel intensely the indignity of our social wrongs; one's very being must throb with the pain, the sorrow, the despair millions of people are daily made to endure. Indeed, unless we have become a part of humanity, we cannot even faintly understand the just indignation that accumulates in a human soul, the burning, surging passion that makes the storm inevitable.
>
> (Goldman 1969: 86)

Goldman believed in revolution, but what that meant was somewhat complicated. She saw revolutionary change as inevitable, but not to be brought about or directed by a self-appointed vanguard: 'Revolution in society is exactly what a cloudburst, an earthquake or a tornado is in the atmospheric realm.' She was not a pure pacifist, advocating 'violence as necessary to defend the revolution' (Ferguson 2011: 296), but she was nevertheless reserved about violence. As she wrote in a 1938 letter during the Spanish Civil War, 'The function of anarchism in a revolutionary period is to minimize the violence of revolution and replace it by constructive efforts' (cited in Porter 1983: 236). In another letter earlier the same year, Goldman reflected on the use of force to bring about change:

> More and more I come to the conclusion that there can be no Anarchist Revolution. By its very violent nature Revolution denies everything Anarchism stands for. The individual ceases to exist, all his rights and liberties go under. In fact life itself becomes cheap and dehumanized.
>
> (cited in Porter 1983: 234)

In this anarchist frame of politics, the urgent, revolutionary tasks are both resistant and creative, and the field of struggle is everywhere and anywhere. This was evident for Goldman not only in the Spanish Revolution, but also in the anarchist movement in which she circulated in New York, where politics were embedded in art and play (and vice versa), and not separate from everyday, ordinary life (see Ferguson 2011: 81). In this world, '"[i]deology, pleasure, and identity" were interwoven' (Goyens 2007: 168, cited in Ferguson 2011: 83). The banal spaces and acts of everyday life are a fundamental political field, some of which is beyond the reach of even a progressive State to regulate. The law and the police cannot decolonize, nor dismantle patriarchy or white supremacy; this can only be accomplished fully through the transformation of social relationships (Goldman 1969; Landauer 2010).

As we pursue a critical urban theory for the inhabited city, we need to think through how everyday life in the city specifically contributes to the oppression of its residents, and how those experiences are differentiated along lines of identity – that those differentiated experiences contribute to the relationships that constitute the identities themselves. This includes spatial segregation of racialized and immigrant groups, migrants who lack political rights in the cities in which they live and/or work (Smith and McQuarrie 2012), inaccessible streetscapes, indigenous peoples whose geographies have been renamed, redeveloped, and erased (Nagam 2015; Johnson 2013), the uneven policing of neighbourhoods, the neglect of public infrastructure, the privatization of common spaces, the refusal to accommodate homeless people, and the physical street attacks on marginalized bodies as individual yet impersonal political acts of exclusion and extermination.

We equally need to recognize the resistance of residents and the way the city provides opportunities to generate new spaces and new relationships. Scott praises Jane Jacobs' argument for the need to 'start[] from the 'lived,' vernacular city' (Scott 2012: 43) and acknowledges the valuable and possibly necessary diversity of 'disorderly city' (41), but his emphasis is the built landscape and land use. We also need to emphasize how individuals have found safety, community and solidarity in cities. The city as a refuge has a deep and variegated history for many groups (Boone 1996; Wilkerson 2010; Takaki 1989; Ridgley 2008). Seeking refuge or struggling to heal is not always a choice, nor is the city a complete or unproblematic place of safety, but the ability to find safety or healing on the same street that endangers you at other moments is nonetheless important to acknowledge. From an anarchist or utopian perspective, the goal is to make such acts and spaces theoretically more visible, to build on them, and to increase and expand the spaces of refuge, health and joy.

There is some symmetry between the concrete yet fluid, anarchist character of the city and its invisible, impossible politics. The politics of the city for marginalized communities, for those who do not have the privilege of choosing their interactions with the State (Cohen 2004), relies on a necessary evasion of the State and other governors, and this search is material and spatial. It is a need to escape surveillance, to find physical places of privacy in public, private and third places, as well as the spaces of physical and emotional safety where one can generate belonging, friendship, community, solidarity, satisfaction, pleasure, joy. The identification and enjoyment of such places requires fragmentation and disengagement from others: those who do not understand or who themselves are oppressors.

Critical urban theory in pursuit of a radical politics should include and address all the ways in which the city liberates and oppresses. It must therefore be explicitly anti-racist, feminist, queer, anti-colonial, anti-ableist, anti-normative and anti-essentialist, and, broadly speaking, intersectionalist and anarchist in its encompassing critique of power relationships and citizenship in its many forms. It should diversify the scope of political thinking, but not only through the inclusion of a diversity of individuals, communities and interests, but with the aim to identify and disempower hierarchies, to decolonize. It should overcome assumptions that oppressed and marginalized communities have no theory of their own, or that they need the 'superior' theory of continental philosophers. Anarchist theory here may serve as an overarching frame or sustaining fuel for the recognition of all oppression. Anarchist theory is proposed not as a master theory nor as a final stop, but as a way into exploring oppression in more detail, in its lived, embodied world.

This is the moment where our understanding of invisible, impossible politics requires citizenship theory, to better understand actual practices of governing, to see how they exceed the State – and how the State may not even be involved. If we draw on a diverse archive of lives and acts, and listen to a choir of theorists who speak from invisible and impossible places, we will appreciate the many different forms this takes: deviance, refusal, graffiti, public acts of art, play, marches, occupations. It includes any act that is taken to expose, resist or evade coercive relationships, and to generate safer spaces, better relationships, moments of wellness and joy. The same city that coerces and causes suffering also provides the flexibility, anonymity, and ephemerality within which this resistance and creativity is possible. Contextualizing acts of citizenship in that understanding of the city leads us to a fundamentally different time and scale of politics. The anarchist idea of the 'propaganda of the deed' is not only a violent act, but any prefigurative praxis, which in its example transforms and also provides evidence for the possibility of transformation. In the acts of deviance, refusal and diva citizenship, there is a politics of the here and now that does not privilege (nor exclude) larger scales or particular forms of liberation (see also Gibson-Graham 2006; Cooper 2014), arising from efforts to survive, to resist violence and to heal from the damage it inflicts.

To speak of 'anarchist' time, however, goes beyond the incorporation of everyday life, and the here and now. Ferguson categorizes the temporalities of Goldman's politics: 'crowd time' arising from the particular dynamics of public speaking, the 'ritual time' of commemorations of people and watershed political moments, and 'the ongoing production of public temporality' through routine publications (Ferguson 2011: 98, 107, 101). Even the use and measurement of time is praxis, responsive to experience and meeting needs as they arise. Time is not merely linear; it is lived.

Berardi argues that our sense of time and history have been driven by ideas of constant improvement and expansion: 'we are accustomed to thinking about time in terms of progress, an endless process of growth, and also in terms of perfectibility' (Berardi 2012: 89). This approach to time led to a futurism with a strong temporal linearity, and which understood the present and future as overcoming and crushing the past. The global financial crisis exposed (again) the fantasy of eternal progress and this means 'the future is over, and we are living in a space that is beyond the future' (Berardi 2012: 81). Berardi believes in the possibility of turning away from the idea of perpetual growth, but as it stands, the future, as we have known it, is dead. Reflecting on Berardi's death of the future, Graeber acknowledges 'a dilemma': that while 'a revolutionary Future' feels impossible, to abandon work towards it is untenable.

> In this sense we are forced to live with two very different futures: that which we suspect will actually come to pass…and the Future in the old revolutionary, apocalyptic sense of the term: the fulfillment of time, the unraveling of contradictions. Genuine knowledge of this Future is

> impossible, but it is only from the perspective of this unknowable Outside that any real knowledge of the present is possible. The Future has become our Dreamtime.
>
> (Graeber 2011: 103)

Those arguing for a radicalization of democracy have also moved away from definitive and deterministic ideas of history and time. Mouffe argues that 'a final resolution of conflicts' is not an appropriate horizon, even if 'asymptotic' (Mouffe 1993: 8). The focus must remain on the ongoing creation of 'an economy that works as if people mattered, and a society at the service of human values and the pursuit of personal happiness' (Castells 2012: 245). Purcell cautions against holding up democracy as 'some end of history we should expect to reach', and similarly emphasizes that 'perpetual democratization is the best way forward in the current context and for the foreseeable future' (Purcell 2013: 3).

Cooper's utopianism also challenges a singular, linear understanding of time, which does not rule out thinking of the future: 'however [everyday utopias] also, importantly, pluralize and complicate this forward direction, challenging as they do so the notion that change lies *only* ahead'. Instead, Cooper writes of the 'complexity and multidirectionality of change…' (Cooper 2014: 218). Utopian work is both aspirational and prefigurative, looking towards the future. But it may also be rooted firmly in the present, which may in turn have a view to the past, 'identifying here the seeds or kernels of something that is (still) to happen, as Marx (2001 [1871]: 84) famously commented in his address on the Paris Commune: 'The workers…have no ideals to realise, but to set free the elements of the new society with which old collapsing bourgeois society itself is pregnant' (Cooper 2014: 219). What is already here? What has already begun?

For Goldman, anarchism and revolution were 'a process, not a finality – "finalities are for gods and governments, not for the human intellect"' (Ferguson 2011: 129, citing Goldman 1972: 49). Time here is a constant flow, shaped but not anchored by history, and its consequent politics is similarly non-linear and asymptotic. We imperfect human beings live in a world in which so many other actors have agency or autonomy. Even with an achievement of 'the perfect order' (whatever that might be, if there is such a thing), it would be inhabited by people who would continue to struggle with themselves and with others, and with their need to separate as well as join. The work of transformation is constant. The non-human forces of the world (plants, animals, climate, plate tectonics) will continue to exercise their own agency, and therefore challenge human desire and undo human accomplishment. The dynamism is not going to reach an end point. Even with agreement on goals or process, there is never only one singular potential outcome. The circumstances are going to create new problems, and demand new arrangements, new relationships, new material and human resources. This imperfection and incompleteness does not contradict anarchist or utopian thinking, both of which are grounded in the reality of here and now as their starting place. While the imagination is an important driver of utopian ideas, they are grounded in the materiality of what already exists – its constraints and potential (Cooper 2014).

Powerful discourse presents capitalism as not only superior, but pervasive and inescapable, yet this is more belief than reality (Gibson-Graham 2006; White and Williams 2014). A diversity of economic forms already exists, where people meet their material and service wants and needs though relationships outside of capitalist market exchange (Gibson-Graham 2006). There are many examples of smaller-scale transformations that have moved beyond experimental prefigurative practices (Gibson-Graham 1996, 2006; Araujo 2016). This way of thinking and the spaces it produces are often framed as 'alternative', but this perspective should be challenged. White and Williams (2014, 2016) argue that a substantial portion of our lives already function this way, that neoliberal capitalism does not saturate our lives to the extent its proponents and critics purport. Some of the distortion arises from definitions of work that privilege paid employment; studies measuring time spent 'at work', defined more broadly, have found a majority of hours engaged in unpaid (or alternative wage) tasks (Burns et al. 2004). The productive economy exceeds the market.

Moreover, non-monetized exchanges have a different character than exploitative capitalist exchange. Rather than individual self-interest and profit, exchanges are motivated by and reproduce care, trust, reciprocity, and solidarity (White 2009). For similar reasons and purposes, 'alternatives' also exist for other-than-economic purposes, finding or creating social spaces beyond the reach of oppressive, coercive or exploitative relationships, including relationships with the State, co-workers, community and family. 'Alternatives' are already commonplace, even dominant in some areas, suggesting that the more-just future we seek is neither impossible nor dead, but already woven through the present.

This approach expands past, present and future beyond linear chronologies. Graeber suggests seeing 'history as a field of permanent possibility', and thus the present is always potentially seeded with transformative ideas and practices that may build the desired future (Graeber 2011: 105; see also Springer 2010). Throughout these rich understandings of time is the importance of the potential, of conditions that are shaped by history without allowing the weight of history to determine their meaning in the present or future. The present and future are always contingent on human imagination and intervention, for

> potential draws (on) a complex and contingent relationship between temporalities, as it also draws (on) a complicated relationship between what is actualized and what is imagined. Potential may inhere within the present, but only because a future is imagined and brought to bear on imagining and guiding what is manifested in the now.
>
> (Cooper 2014: 220)

The constitutionalism of invisible, impossible politics

All this work of resistance and generation must be made visible, and should be the basis for critical urban theory. Moreover, this work is the basis for imagining a possible 'people' at any scale, not to be united or represented, but to be constituted as citizens; this work is a higher order of politics. It is a form of (a) revolution and (b) constitutionalism, because it acts to change the principles we live by, which thus constitute 'the people' as a polity living together non-coercively.

The term 'revolution' should be understood, or perhaps *recognized*, on different time and spatial scales, in its different manifestations. Two examples are useful for illustrating this, one 'classic' and one which might be deemed 'alternative'. The French Revolution is commonly understood as the overthrow from a totalitarian form of State (monarchy), as well as the overthrow of the social governance of the nobility and the Church, by 'the people', in favour of democratic forms of governance – again, both political and social. Historians and others have struggled mightily with the apparent contradictions of the Terror that came in the wake of the Revolution, wondering if it was a necessary part of a transition to democracy (see, for example, Andress 2006; Tackett 2015). Certainly at the time, advocates of democracy such as Edmund Burke were appalled at the Terror (Burke 1790), and the felt need to avoid any repetition of such violence and societal breakdown has shaped theoretical definitions and the actual implementation of democratic institutions since. Rancière argues, in his reading of Francois Furet's 1978 *Interpreting the French Revolution*, that

> the revolutionary theatre in its entirety was founded on an ignorance of the profound historical realities that rendered it possible. It ignored the fact that the true revolution, that of institutions and moral values, had already taken place in the depths of society and in the cogs of the monarchical machine.
>
> (Rancière 2014: 15)

This is significant not only for the debates regarding the meaning and purpose of the Terror, but also for our determination as to what constitutes the R/revolution itself. By what means was change brought about, and at what point do we determine that it was revolutionary? Is there a privileging of formal political structure in our understanding of revolution, and is it appropriate or productive? More fundamentally, is there a privileging of political structure in our understanding of governance? Whose lives are consequentially privileged by such biases? Remembering the accusation of O'Casey's Mrs Madigan, in a revolution whose life is changed and how? Rancière's language of 'true revolution, that of institutions and moral values' indicates his position that the form of the State is secondary, to the extent it is relevant at all.

It is worth underscoring (again) that this critique is not an anachronistic application of twenty-first-century Western values (whatever they might be). Olympe de Gouges answered this at the time in her 1791 'Declaration of the Rights of Woman and the Female Citizen', when she pointed out that women's emancipation also lay within social relationships and institutions, not political structures, although it was within the reach of a legislative assembly to address them in part. Specifically, she argued for the recognition of women's work bearing and raising children, and of women's unequal position within the institution of marriage as it was legally structured at the time. She argued that revolutionaries, if they were committed to the principles of democracy and equality they claimed to espouse, were obliged to follow the logic through to the end. And she called them out for not doing so. This and other writings were deemed an attack on the Revolutionary Government and Olympe de Gouges was guillotined in 1793 by the Jacobins.

An equally profound yet less violent example worth comparing is the 'Quiet Revolution' in Quebec. Over the course of two or three decades following the Second World War, there were deep changes that became the foundation for the ascendancy of (nominally) Catholic, French-speaking residents to challenge the social, economic, cultural, linguistic and political dominance of American corporations and the Anglo-Canadian minority in the province. Not coincidentally, this arose in part out of the urbanization of the French-speaking Quebec population (Behiels 1985). This revolution never overthrew the State, and indeed, many worked within the existing political system at the provincial level, and then at the federal level, to create new political parties to assert their voice. But through a programme of institutional and social secularization, they did 'overthrow' the Church, which in so many ways had functioned as a state, governing their lives, publicly and privately, through houses of religion but also hospitals, schools, and a myriad of moralizing social practices and discourses. The Quiet Revolution also reasserted the sovereignty of francophones whose language had been pushed to the margins of public life. Anglophone political leaders were replaced, but more profoundly, French was reestablished as the public language of politics, business and daily life in the province's most powerful city of Montreal. Colonizing relationships were undone. Mindful of intersectionality, however, it is important to note that the separatist Québecois nationalism that arose from the Quiet Revolution is not unproblematic, as it remains in tension with francophone Muslim immigrants and refugees (Gilbert 2009; Bouchard 2015), as well as indigenous rights and land claims within what is claimed as provincial territory (Desbiens 2004).

On a *much* more minor scale, the Quiet Revolution was accompanied by public violence from the sovereigntist paramilitary group, the *Front de Libération du Québec* (FLQ). Beginning in 1963, they were responsible for a series of bombings that caused several injuries and deaths, including bombing the Montreal Stock Exchange and the house of the mayor of Montreal in 1969. Their campaign culminated in the October Crisis of 1970, with two political kidnappings, one of which ended in the death of the hostage, the provincial

Labour Minister, Pierre Laporte. The Crisis was the basis for the federal government's invocation of the War Measures Act (by Prime Minister Pierre Trudeau, a leading intellectual of the Quiet Revolution), which suspends several civil rights, and had never before been employed in peacetime. There is a slight parallel here to the Terror of the French Revolution as Rancière views it, in that the majority of the social and cultural transformations were already complete or well underway by the time the FLQ organized in pursuit of the overthrow of the Quebec government and political independence from Canada, and even before the formal political transformations of the 1960s (Behiels 1985; Gauvreau 2008).

Nepstad (2011) has argued that nonviolent revolutionary change is not only possible, but has a greater success rate, in terms of its durability. Graeber is similarly critical of the idea of an event such as 'a single mass insurrection' as constituting a revolution, raising the example of feminism's (continuing) unravelling of patriarchy over the course of centuries as the very definition of radical change. He believes we need to move away from the idea of revolution as 'insurrectionary', although he admits 'it's not clear what will replace it' (Graeber 2011: 42). The idea of revolution is not singular, nor neatly measured in terms of time. Acts of diva citizenship, of refusal and of creation are revolutionary as they replace or dismantle coercive relationships. If we focus only on institutional change, permanence, or mass mobilization, we miss their success and power, and we fail to recognize the suffering they reduce, however briefly.

These acts of citizenship also contribute to a higher order of politics, constitutionalism. Most of the literature on constitutionalism, like that on democratic theory in general, focuses on formal political institutions and documents. Kalyvas has pointed, however, to the important work done by collective action outside the system, which feeds and shapes formal politics in practice. Informal or 'extraordinary' politics is not necessarily the opposite of formal or 'ordinary' politics; they are in some ways mutually dependent, relying on the differences between them. Efforts to formalize the informal practices are counter-productive, as 'the extraordinary is negated by the urge to turn it into a lasting, that is, an ordinary, phenomenon. Its singular character is lost' (Kalyvas 2008: 298). Kalyvas argues for 'a plurality of social movements and voluntary political associations as the inescapable ground upon which popular sovereignty is constructed'. These movements should not be dismissed for their informality nor misunderstood as a distortion of politics. They are 'self-constituted and self-formed networks and discourses' of politics in their own right. 'As such they testify to the creative capacity of collective actors to develop spontaneous self-organized counterinstitutions apart from the juridical system and the constituted order of the state machine' (Kalyvas 2008: 299).

The impact of extraordinary politics is not limited to legislative change. Isin and Ruppert (2015) have examined a related phenomenon, in the governance of the Internet as well as its use in acts of citizenship. These acts are overwhelmingly in the sphere of the extraordinary, sometimes mapping onto states, sometimes transcending them. 'Digital citizens' must constitute themselves as a kind of polity through the process of making claims on those who control the channels of access to the Internet. This entails new and complex relationships, alliances and negotiations to determine what claims to make and how to do so successfully. Through these acts and 'the imaginary force of these bills, charters, declarations, and manifestos', 'digital citizens' are constituted who participate successfully in formal institutions of governance, including corporations (Isin and Ruppert 2015: 178).

Building on Kalyvas and expanding the extraordinary to include the range of political acts of the inhabited city, I argue that these informal acts expose the citizenship gap and thus challenge the principles that constitute the polity in practice. They dispute the sincerity of formal constitutions and the equalizing promise of citizenship as an institution and as an identity. They accuse the State itself of violating its own principles, of oppressing violently its own citizens. They thus raise questions about the meaning and practice of democracy, and the capacity of the State to deliver on its promises of governing. They demand the State be accountable and constitute themselves as citizens who are entitled to make such a claim.

'The people' cannot be represented by other people, even by mass mobilizations. The people can be represented, indeed constituted, by agreements, written and unwritten, that set out the principles of a polity. The people are usually not the literal signatories of such a document nor the legislators who may have crafted it, but those whom it encompasses. As Isin and Ruppert write,

> Even if it is done in the name of the people, a declaration has no way of guaranteeing that in fact its people exist or will exist as a fact. Rather, and this is Derrida's intervention, the act brings the people, its political subject, into being through the act. The people a declaration names do not exist. Derrida writes, '[People] do *not* exist, *before* this declaration, not as such.'
>
> (Isin and Ruppert 2015: 177)

'The people' are citizens.

These constitutions by which we live are multiple, covering the many forms in which citizens are governed: by the State, by corporations, by each other. Written or unwritten, they are living, evolving ideas.

All forms of resistance must be recognized, and we should resist categorizing them as revolutionary or not, as a revolution does not occur with one, single act, and we can have no idea of the ripple effects of any act. As Scott underscored, a political action may later have great import, unknown at its spontaneous eruption:

> Revolutions and social movements are, then, typically confected by a plurality of actors: actors with wildly divergent objectives mixed with a

> large dose of rage and indignation, actors with little knowledge of the situation beyond their immediate ken, actors subject to chance occurrences…and yet the vector sum of this cacophony of events may set the stage for what later is seen as a revolution. They are rarely, if ever, the work of coherent organizations directing their 'troops' to a determined objective, as the Leninist script would have it.
>
> (Scott 2012: 138–139)

Individuals who gathered in squares and parks in cities that did not accommodate them, who generated temporarily a non-coercive space of exchange and struggle, who sang and drummed, who read manifestoes or didn't – all these forged another link in a chain of urban practices of resistance that preceded them by generations. The nature of their political acts reflects their lived experiences and the structures of their cities. They held in common their anger, hope and ephemeral presence with other occupiers and marchers who came before them, but also the book-burners, the performance artists, the stone-throwers, the graffiti artists and flash mob dancers. These were not formal organizations; many occupations did not lead to institutional change or even continued solidarity of its participants. And yet, their political contributions were real and substantive; they confronted their communities, themselves and each other with the many failures to enact in practice the values they preach.

The difference between Marxian and anarchist approaches that I am underscoring is not their positions on the role of the State, but what I consider a more abstract and fundamental difference in the understanding of what constitutes politics and its temporal and spatial frames. The politics of the invisible challenge our ideas of what is impossible: they challenge the constitution of our cities and our polities, the principles by which we live. This diversity and radicalism must be the foundation of urban critical theory and a broader radical politics of the Left. It must be intersectionalist, flexible, grounded in praxis as much as abstract theory, acknowledging the hatred and intentionality that lies behind the 'second-class' citizenship so many are subjected to. It must acknowledge the subsequent generations of suffering and trauma, and seek not only to achieve justice, but also to heal.

Note

1 Scott has explored anarchist theory more explicitly elsewhere (see Scott 2009, 2012).

References

Aalbers, M. 2008. 'The Financialisation of Home and the Mortgage Market Crisis', *Competition and Change*, 12:2, 148–166.

Acevedo, Elizabeth. 2016. 'Split This Rock Interview with Dominique Christina', *Split This Rock*, 30 March. http://blogthisrock.blogspot.ca/2016/03/split-this-rock-interview-with.html

Adewunmi, Bim. 2013. 'Artist-Activist Liz Crow's "Bed-Out": For Disabled Rights', *The Guardian*, 10 April. http://www.theguardian.com/artanddesign/2013/apr/09/liz-crow-bed-disabled-rights

Ahmed, Sara. 2004. *The Cultural Politics of Emotion*, London: Routledge.

Amit-Talai, Vered and Henri Lustiger-Thaler (eds) 1994. *Urban Lives: Fragmentation and Resistance*, Toronto: McClelland and Stewart.

Alfred, G.R. 1995. *Heeding the Voices of Our Ancestors: Kahnawake Mohawk Politics and the Rise of Native Nationalism*, Don Mills, ON: Oxford University Press.

Alfred, Taiaike. 2005. *Wasáse: Indigenous Pathways of Action and Freedom*, Peterborough: Broadview Press.

Andress, David. 2006. *The Terror: The Merciless War for Freedom in Revolutionary France*, New York: Farrar, Straus and Giroux.

Araujo, Erin. 2016. 'Consensus and Activism through Collective Exchanges: A Focus on El Cambalache, Mexico', *International Journal of Sociology and Social Policy*, 36:11/12, 741–755.

Arbaci, Sonia and Jorge Malheiros. 2010. 'De-segregation, Peripheralisation and the Social Exclusion of Immigrants: Southern European Cities in the 1990s', *Journal of Ethnic and Migration Studies*, 36:2, 227–255.

Bain, Alison L. 2004. 'In/visible Geographies: Absence, Emergence, Presence, and the Fine Art of Identity Construction', *Tijdschrift voor Economische en Sociale Geografie*, 95:4, 419–426.

Balzer, Samantha. 2014. 'Beginning with the Body: Fleshy Politics in the Performance Art of Rebecca Belmore and Leah Lakshmi Piepzna-Smarasinha', *Journal of Feminist Scholarship*, 6, 47–58.

Bamyeh, Mohammed A. 2012. 'The Global Culture of Protest', *Contexts*, 11:2, Spring, 16–18.

Barber, Benjamin. 2012. 'What Democracy Looks Like', *Contexts*, 11:2, Spring, 14–16.

Battersby, Sarah-Joyce. 2016. 'Inside Toronto's Black Lives Matter Camp', *Toronto Star*, 3 April. http://www.thestar.com/news/gta/2016/04/03/inside-torontos-black-lives-matter-camp.html

Behiels, Michael D. 1985. *Prelude to Quebec's Quiet Revolution: Liberalism vs Neo-Nationalism, 1945–60*, Montreal and Kingston: McGill-Queen's University Press.

Berardi, Franco 'Bifo'. 2012. *The Uprising: On Poetry and Finance*, Los Angeles: Semiotext(e).

Berlant, Lauren. 1997. *The Queen of America Goes to Washington City: Essays on Sex and Citizenship*, Durham, NC: Duke University Press.

Bicchieri, Cristina. 2006. *The Grammar of Society: The Nature and Dynamics of Social Norms*, Cambridge: Cambridge University Press.

Bissonette, Devan. 2014. 'A Digital Democracy or Twenty-First-Century Tyranny? CNN's iReport and the Future of Citizenship in Virtual Spaces', in Matt Ratto and Megan Boler (eds), *DIY Citizenship: Critical Making and Social Media*, Cambridge, MA and London: MIT Press, 385–401.

Bofkin, Lee. 2014. *Global Street Art: The Street Artists and Trends Taking Over the World*, London: Cassell.

Boler, Megan. 1999. *Feeling Power: Emotions and Education*, New York: Routledge.

Bonaiuti, Gianluca. 2013. 'La Parte. Note sulla politica del «popolo» in Jacques Rancière', *Meridiana: Rivista di Storia e Scienze Sociali*, 77, 145–174.

Bookchin, Murray. 2015. *The Next Revolution: Popular Assemblies and the Promise of Direct Democracy*, Debbie Bookchin and Blair Taylor (eds), London and New York: Verso.

Boone, J.A. 1996, 'Queer Sites in Modernism: Harlem/the Left Bank/Greenwich Village', in P. Yaeger (ed.), *The Geography of Identity*, Ann Arbor, MI: University of Michigan Press, 243–272.

Bosteels, Bruno. 2010. 'Politics, Infrapolitics, and the Impolitical: Notes on the Thought of Roberto Esposito and Alberto Moreiras', *CR: The New Centennial Review*, 10:2, 205–238.

Bouchard, Gérard. 2015. *Interculturalism: A View from Quebec*, Toronto: University of Toronto Press.

Breckenridge, Carol A. and Candace Volger. 2001. 'The Critical Limits of Embodiment: Disability's Criticism', *Public Culture*, 13:3, 349–357.

Brenner, Neil. 2011. 'What Is Critical Urban Theory?' in Neil Brenner, Peter Marcuse and Margit Mayer (eds), *Cities for People, Not for Profit: Critical Urban Theory and the Right to the City*, London and New York: Routledge, 10–23.

Brenner, Neil (ed.) 2014. *Implosions/Explosions: Towards a Study of Planetary Urbanization*, Berlin: Jovis Verlag.

Brenner, Neil and Christian Schmid. 2012. 'Planetary Urbanization', in Mathew Gandy (ed.), *Urban Constellations*. Berlin: Jovis Verlag, 10–13.

Brenner, Neil, David J. Madden and David Wachsmuth. 2011. 'Assemblages, Actor-Networks, and the Challenges of Critical Urban Theory', in Neil Brenner, Peter Marcuse and Margit Mayer (eds), *Cities for People, Not for Profit: Critical Urban Theory and the Right to the City*, London and New York: Routledge, 117–137.

Brenner, Neil, Peter Marcuse and Margit Mayer. 2011. 'Cities for People, Not for Profit: An Introduction', in Neil Brenner, Peter Marcuse and Margit Mayer (eds), *Cities for People, Not for Profit: Critical Urban Theory and the Right to the City*, London and New York: Routledge, 1–9.

Brown, Michael P. 1995. *RePlacing Citizenship: AIDS Activism and Radical Democracy*, New York: Guilford Press.

Brunson, Rodney K. 2007. '"Police don't like black people": African American Young Men's Accumulated Police Experiences', *Criminology & Public Policy*, 61, 71–102.

Burke, Edmund. 1790. *Reflections on the Revolution in France*, London: J. Dodsley.
Burns, D., Colin C. Williams and J. Windebank. 2004. *Community Self-help*, Basingstoke: Palgrave Macmillan.
Butler, Judith. 1998. 'Merely Cultural', *New Left Review*, 0.227 (January–February), 33–44.
Byrne, Janet (ed.) 2012. *The Occupy Handbook*, New York: Little, Brown.
Castells, Manuel. 1983. *The City and the Grass Roots: A Cross-Cultural Theory of Urban Social Movements*, Berkeley: University of California Press.
Castells, Manuel. 2012. *Networks of Outrage and Hope: Social Movements in the Internet Age*, Cambridge and Malden, MA: Polity Press.
Castells, Manuel, João Caraça and Gustavo Cardoso. 2012. 'The Cultures of the Economic Crisis: An Introduction', in Manuel Castells, João Caraça and Gustavo Cardoso (eds), *Aftermath: The Cultures of the Economic Crisis*, Oxford: Oxford University Press, 1–14.
Chatterton, Paul and Jenny Pickerill. 2010. 'Everyday Activism and Transitions towards Post-capitalist Worlds', *Transactions of the Institute of British Geographers*, 35:2, 475–490.
Chauncey, G. 1995. *Gay New York: Gender, Urban Culture and the Making of the Gay Male World, 1890–1940*, New York: Basic Books.
Chomsky, Noam. 2012. *Occupy* (Occupied Media Pamphlet Series), New York: Zuccotti Park Press.
Christian, Barbara. 1987. 'The Race for Theory', *Feminist Studies*, 14:1, 67–79.
Christina, Dominique. 2015. *This is Woman's Work*, Boulder, CO: Sounds True.
Christophers, Brett. 2011. 'Revisiting the Urbanization of Capital', *Annals of the Association of American Geographers*, 101, 1347–1364.
Cohen, Cathy J. 2004. 'Deviance as Resistance: A New Research Agenda for the Study of Black Politics', *Du Bois Review*, 1: 27–45.
Cohen, Cathy J. 2010. *Democracy Remixed: Black Youth and the Future of American Politics*, New York: Oxford University Press.
Cole, P. 2006. *Coyote and Raven Go Canoeing: Coming Home to the Village*, Montreal, QC: McGill-Queen's University Press.
Collins, Patricia Hill. 1990. *Black Feminist Thought: Knowledge, Consciousness and the Politics of Empowerment*, London: Routledge.
Collins, Patricia Hill. 1998. *Fighting Words: Black Women and the Search for Justice*, Minneapolis: University of Minnesota Press.
Colliot-Thélène, Catherine. 2011. *La Démocratie sans «Demos»*, Paris: PUF.
Contexts editors. 2012. 'Understanding "Occupy"', *Contexts*, 11:2, Spring, 12–13.
Cooper, Davina. 1998. *Governing Out of Order: Space, Law and the Politics of Belonging*, London and New York: Rivers Oram Press.
Cooper, Davina. 2014. *Everyday Utopias: The Conceptual Life of Promising Spaces*, Durham, NC and London: Duke University Press.
Council of Europe. 2013. *Human Rights of Travellers and Roma in Europe*. Available at https://book.coe.int/eur/en/minorities/5706-e-pub-human-rights-of-roma-and-travellers-in-europe.html
Cox, Laurence and Alf Gunvald Nilsen. 2014. *We Make Our Own History: Marxism and Social Movements in the Twilight of Neoliberalism*, London: Pluto Press.
Crenshaw, Kimberlé. 1989. 'Demarginalizing the Intersection of Race and Sex', *University of Chicago Legal Forum*, 139–167.
Crenshaw, Kimberlé. 1991. 'Mapping the Margins: Intersectionality, Identity Politics, and Violence against Women of Color', *Stanford Law Review*, 43:6, 1241–1299.

Crowley, Una. 2005. 'Liberal Rule through Non-Liberal Means: The Attempted Settlement of Irish Travellers (1955–1975)', *Irish Geography*, 38:2, 128–150.
Crowley, Una. 2009. 'Outside in Dublin: Travellers, Society and the State, 1963–1985', *The Canadian Journal of Irish Studies*, 35:1, 17–24.
Deitsch, Jeffrey, with Roger Gastman and Aaron Rose. 2011. *Art in the Streets*, New York: Skira Rizzoli Publications.
Derickson, Kate Driscoll. 2016. 'On the Politics of Recognition in Critical Urban Scholarship', *Urban Geography*, 37:6, 824–839.
Desbiens, Caroline. 2004. 'Nation to Nation: Defining New Structures of Development in Northern Quebec', *Economic Geography*, 80:4, 351–366.
Deutsche, Rosalyn. 1996. *Evictions: Art and Spatial Politics*, Cambridge, MA: MIT Press.
Dowler, Lorraine. 2001. 'No Man's Land: Gender and the Geopolitics of Mobility in West Belfast, Northern Ireland', *Geopolitics*, 6:3, 158–176.
Duso, Giuseppe. 2010. 'Thinking about Politics beyond Modern Concepts', Stephen Marth (trans.), *CR: The New Centennial Review*, 10:2, 73–97.
Emmanuele, Vincent. 2015. 'Poetic Resistance: An Interview with Dominique Christina', *Countercurrents.org*, 17 February. http://www.countercurrents.org/emanuele170215.htm
Fallon, Amy. 2015. 'Gay Ugandans Launch Magazine to "Reclaim Stories"', *Yahoo News*, 9 February. https://www.yahoo.com/news/gay-ugandans-launch-magazine-reclaim-stories-060203422.html?ref=gs
Ferguson, Kathy E. 2011. *Emma Goldman: Political Thinking in the Streets*, London: Rowman & Littlefield.
Fraser, Nancy. 1997a. *Justice Interruptus: Critical Reflections on the 'Postsocialist' Condition*, New York and London: Routledge.
Fraser, Nancy. 1997b. 'A Rejoinder to Iris Young', *New Left Review*, 0.223 (May–June), 126–129.
Fraser, Nancy. 1998. 'Response to Judith Butler', *New Left Review*, 0.228 (March–April), 140–149.
Fukuyama, Francis. 1992. *The End of History and the Last Man*, New York: Free Press.
Gauvreau, Michael. 2008. *The Catholic Origins of Quebec's Quiet Revolution, 1931–1970*, Montreal and Kingston: McGill-Queen's University Press.
Georgakas, Dan and Marvin Surkin. 2012. *Detroit: I Do Mind Dying: A Study in Urban Revolution*, 3rd edn, Chicago: Haymarket Books.
Gerbaudo, Paolo. 2012. *Tweets and the Streets: Social Media and Contemporary Activism*, London: Pluto Press.
Gessen, Keith, Sarah Leonard, Carla Blumenkranz, Mark Greif, Astra Taylor, Sarah Resnick, Nikil Saval and Eli Schmitt (eds) 2011. *Occupy! Scenes from Occupied America*, New York: Verso.
Gibson, Katherine. 2014. 'Thinking Around What a Radical Geography "Must Be"', *Dialogues in Human Geography*, 4:3, 283–287.
Gibson-Graham, J.K. 1996. *The End of Capitalism (as we knew it): A Feminist Critique of Political Economy*, London: Wiley-Blackwell.
Gibson-Graham, J.K. 2006. *A Postcapitalist Politics*, Minneapolis and London: University of Minnesota Press.
Gieseking, Jen Jack. 2016. 'Crossing Over into Neighbourhoods of the Body: Urban Territories, Borders and Lesbian-Queer Bodies in New York City', *AREA*, 48:3, 262–270.

Gilbert, Liette. 2009. 'Immigration as Local Politics: Re-Bordering Immigration and Multiculturalism through Deterrence and Incapacitation', *International Journal of Urban and Regional Research*, 33:1, 26–42.
Gilbert, Sandra and Susan Gubar. 1979. *The Madwoman in the Attic: The Woman Writer and the Nineteenth-Century Literary Imagination*, New Haven, CT and London: Yale University Press.
Giordano, J.M. 2016. 'Street Artists Bring Political Murals to Baltimore – in Pictures', *The Guardian*, 19 May. Available at: https://www.theguardian.com/artanddesign/gallery/2016/may/19/street-artists-bring-political-murals-to-baltimore-in-pictures
Gitlin, Todd. 2012. *Occupy Nation: The Roots, the Spirit, and the Promise of Occupy Wall Street*, New York: It Books.
Goldman, Emma 1969 [1910]. *Anarchism and Other Essays*, Port Washington, NY: Kennikat Press.
Goldman, Emma. 1972. *Red Emma Speaks: Selected Writing and Speeches*, Alix Kates Shulman (ed.), NewYork: Vintage Books.
Goonewardena, Kanishka. 2011. 'Space and Revolution in Theory and Practice: Eight Theses', in Neil Brenner, Peter Marcuse and Margit Mayer (eds), *Cities for People, Not for Profit: Critical Urban Theory and the Right to the City*, London and New York: Routledge, 86–101.
Goonewardena, K., S. Kipfer, R. Milgrom and C. Schmid (eds) 2008. *Space, Difference, Everyday Life: Reading Henri Lefebvre*, New York: Routledge.
Gotham, K.F. 2009. 'Creating Liquidity out of Spatial Fixity: The Secondary Circuit of Capital and the Subprime Mortgage Crisis', *International Journal of Urban and Regional Research*, 332, 355–371.
Gough, Jamie. 2002. 'Neoliberalism and Socialisation in the Contemporary City: Opposites, Complements and Instabilities', *Antipode*, 34:3, 405–426.
Gould, Deborah B. 2012. 'Occupy's Political Emotions', *Contexts*, 11:2, Spring, 20–21.
Goyens, Tom. 2007. *Beer and Revolution: The German Anarchist Movement in New York City, 1880–1914*, Urbana: University of Illinois Press.
Graeber, David. 2011. *Revolutions in Reverse: Essays on Politics, Violence, Art, and Imagination*, London and New York: Autonomedia.
Graeber, David. 2013. *The Democracy Project*, New York: Speigel and Grau.
Graham, Stephen and Simon Marvin. 2001. *Splintering Urbanism, Networked Infrastructures, Technological Mobilities and the Urban Condition*, New York: Routledge.
Greenberg, Cheryl. 1992. 'The Politics of Disorder: Reexamining Harlem's Riots of 1935 and 1943', *Journal of Urban History*, 184, 395–441.
Grubbs, David. 2014. *Records Ruin the Landscape: John Cage, the Sixties, and Sound Recording*, Durham, NC: Duke University Press.
Hackworth, Jason. 2006*The Neoliberal City: Governance, Ideology, and Development in American Urbanism*, Ithaca, NY: Cornell University Press.
Harcourt, Bernard. 2013. 'Political Disobedience', in W.J.T. Mitchell, Bernard Harcourt and Michael Taussig, *Occupy: Three Inquiries in Disobedience*, Chicago: TRIOS/University of Chicago Press, 45–92.
Hardt, Michael and Antonio Negri. 2004. *Multitude: War and Democracy in the Age of Empire*, New York: Penguin.
Harrison, Yannick N.A. and Bjarke S. Risager. 2016. 'Convergence and Organization: Blockupy against the ECB', *International Journal of Sociology and Social Policy*, 36:11/12, 843–859.

Harvey, David. 1985. *The Urbanization of Capital: Studies in the History and Theory of Capitalist Urbanization*, Baltimore, MD: Johns Hopkins University Press.

Harvey, David. 2013. *Rebel Cities: From the Right to the City to the Urban Revolution*, London: Verso.

Harvey, David (with David Wachsmuth). 2011. 'What is to Be Done? And Who the Hell is Going to Do It?' in Neil Brenner, Peter Marcuse and Margit Mayer (eds), *Cities for People, Not for Profit: Critical Urban Theory and the Right to the City*, London and New York: Routledge, 264–274.

Harvey, David. 2017. 'Neoliberalism is a Political Project', *Jacobinmag.com*. 23 July. Available at https://www.jacobinmag.com/2016/07/david-harvey-neoliberalism-capitalism-labor-crisis-resistance/

Harvey, David. 2016b. [in press] 'Listen, Anarchist!' *Dialogues in Human Geography*.

Hearne, Rory. 2015. 'The Silent Destruction of Community Growth', *Irish Examiner*. 29 July.

Heathcott, J. 2005. 'Black Archipelago: Politics and City Life in the Jim Crow City', *Journal of Social History*, 38, 705–736.

Hedva, Johanna. 2015. 'Sick Woman Theory', adapted from the lecture, 'My Body Is a Prison of Pain so I Want to Leave It Like a Mystic But I Also Love It & Want It to Matter Politically', delivered at Human Resources, sponsored by the Women's Center for Creative Work, in Los Angeles, on 7 October, *Mask Magazine*, http://www.maskmagazine.com/not-again/struggle/sick-woman-theory

Herman, Edward and Noam Chomsky. 1988. *Manufacturing Consent: The Political Economy of the Mass Media*, New York: Pantheon Books.

Hogg, Alec. 2015. 'As Inequality Soars, the Nervous Super Rich are Already Planning their Escapes', *The Guardian*, 23 January. https://www.theguardian.com/public-leaders-network/2015/jan/23/nervous-super-rich-planning-escapes-davos-2015

Hohle, Randolph. 2015. *Race and the Origins of American Neoliberalism*, New York: Routledge.

Holston, James. 1998. 'Spaces of Insurgent Citizenship', in L. Sandercock (ed.), *Making the Invisible Visible: A Multicultural Planning History*, Berkeley, CA: University of California Press, 37–56.

Holston, James. 2008. *Insurgent Citizenship*. Princeton, NJ: Princeton University Press.

hooks, bell. 1981. *Ain't I a Woman: Black Women and Feminism*, Boston: South End Press.

hooks, bell. 1991. *Yearning: Race, Gender and Cultural Politics*, London: Turnaround.

Huizinga, Johan. 1970. *Homo Ludens: A Study of the Play Element in Culture*, London: Temple Smith.

Hull, Gloria T., Patricia Bell Scott and Barbara Smith. 1982. *All the Women Are White, All the Blacks Are Men, But Some of Us Are Brave: Black Women's Studies*, New York: The Feminist Press at the City University of New York.

Iceland, John. 2009. *Where We Live Now: Immigration and Race in the United States*, Berkeley, CA: University of California Press.

Isin, Engin F. 2002. *Being Political: Genealogies of Citizenship*, Minneapolis, MN: University of Minnesota Press.

Isin, Engin F. and Greg M. Nielsen (eds). 2008. *Acts of Citizenship*, London and New York: Zed Books.

Isin, Engin F. and Evelyn Ruppert. 2015. *Being Digital Citizens*, London: Rowman & Littlefield.

Isin, Engin F. and Patricia Wood. 1999. *Citizenship and Identity*, London: Sage.

Jarrett, Kylie. 2015. *Feminism, Labour and Digital Media: The Digital Housewife*, New York and London: Routledge.
Johnson, Jon. 2013. 'The Great Indian Bus Tour: Mapping Toronto's Urban First Nations Oral Tradition', in Karl Hele (ed.), *The Nature of Empires and the Empires of Nature: Indigenous Peoples and the Great Lakes Environment*, Waterloo, ON: Wilfrid Laurier University Press, 279–297.
Johnson, Theodore R. 2014. 'Don't Let Ferguson Fizzle Like Occupy', *The Root*, 1 December. http://www.theroot.com/articles/culture/2014/12/don_t_let_ferguson_fizzle_like_occupy/
Kalyvas, Andreas. 2008. *Democracy and the Politics of the Extraordinary: Max Weber, Carl Schmitt, and Hannah Arendt*, New York: Cambridge University Press.
Katz, Cyndi. 2006. 'Messing with "the Project"', in Noel Castree and Derek Gregory (eds), *David Harvey: A Critical Reader*, Malden, MA: Blackwell, 234–246.
Kelley, Blair L.M. 2010. *Right to Ride: Streetcar Boycotts and African American Citizenship in the Era of Plessy v. Ferguson*, Chapel Hill, NC: University of North Carolina Press.
Kelley, Blair L.M. 2014. 'The History Behind that "Angry Black Woman" Riff the NY Times Tossed Around', *TheRoot.com*, 25 September. http://www.theroot.com/articles/culture/2014/09/the_angry_black_woman_stereotype_s_long_history.html
King, Thomas. 1990. 'Godzilla vs the Postcolonial', *World Literature Written in English*, 30:2, 10–16.
King, Thomas. 2003. *The Truth about Stories: A Native Narrative*, Toronto: House of Anansi Press.
Klodawsky, Fran, Caroline Andrew and Janet Siltanen. 2016. '"The City Will Be Ours: We Have So Decided": Circulating Knowledge in a Feminist Register', *ACME: An International Journal for Critical Geographies*, 15:2, 309–329.
Knadler, Stephen. 2013. 'Disabled Citizenship: Narrating the Extraordinary Body in Racial Uplift', *Arizona Quarterly: A Journal of American Literature, Culture and Theory*, 69:3, 99–128.
Laclau, Ernesto and Chantal Mouffe. 2001 [1985]. *Hegemony & Socialist Strategy: Towards a Radical Democratic Politics*, 2nd edn, London: Verso.
Lakoff, George and Mark Johnson. 2003. *Metaphors We Live By*, Chicago: University of Chicago Press.
Landauer, Gustav. 2010. *Revolution and Other Writings: A Political Reader*, Gabriel Kuhn (ed. and trans.), Oakland: PM Press.
Lees, L., Tom Slater and Elvin Wyly. 2008. *Gentrification*, New York and London: Routledge.
Lefebvre, Henri. 1970. *La révolution urbaine*, Paris: Gallimard.
Lefebvre, Henri. 1991 [1974]. *The Production of Space*, D. Nicholson-Smith (trans.), Oxford: Blackwell.
Lefebvre, Henri. 1996. 'The Right to the City', in Eleanor Kofman and E. Lebas (ed. and trans.), *Writings on Cities*, Oxford and New York: Wiley-Blackwell.
Leitner, Helga, Eric Sheppard and Kristin M. Sziarto, 2008. 'The Spatialities of Contentious Politics', *Transactions of the Institute of British Geographers*, 33: 157–172.
Leslie, Keith. 2016. 'Grassy Narrows Protesters Bring Mercury Fears to the Legislature', *Toronto Star*, 23 June. https://www.thestar.com/news/queenspark/2016/06/23/grassy-narrows-protesters-bring-mercury-fears-to-the-legislature.html
Lister, Ruth. 1998. *Citizenship: Feminist Perspectives*, New York: NYU Press.

Lister, Ruth. 2007. '(Mis)recognition, Social Inequality and Social Justice: A Critical Social Policy Perspective', in Terry Lovell (ed.), *(Mis)recognition, Social Inequality and Social Justice: Nancy Fraser and Pierre Bourdieu*, London and New York: Routledge, 157–176.

Longmore, Paul K. 2003. *Why I Burned My Book and Other Essays on Disability*, Philadelphia: Temple University Press.

Lorde, Audre. 1984. *Sister Outsider: Essays and Speeches*, Trumansburg, NY: Crossing Press.

Lorde, Audre. 1988. *A Burst of Light: Essays*, Ithaca, NY: Firebrand Books.

Lovell, Terry. 2007. 'Introduction', in Terry Lovell (ed.), *(Mis)recognition, Social Inequality and Social Justice: Nancy Fraser and Pierre Bourdieu*, London and New York: Routledge, 1–16.

Lustiger-Thaler, Henri. 1994. 'Community and Social Practices: The Contingency of Everyday Life', in Vered Amit-Talai and Henri Lustiger-Thaler (eds), *Urban Lives: Fragmentation and Resistance*, Toronto: McClelland and Stewart, 20–44.

Ly, Philip. 2016. 'Wheelchair Protesters Call for Change in La Paz', *SBS*, 4 May. http://www.sbs.com.au/news/article/2016/05/04/wheelchair-protesters-call-change-la-paz

MacKinnon, Catharine A. 1989. *Toward a Feminist Theory of the State*, Cambridge, MA: Harvard University Press.

Marcuse, Peter. 2011. 'Whose Right(s) to What City?" in Neil Brenner, Peter Marcuse and Margit Mayer (eds), *Cities for People, Not for Profit: Critical Urban Theory and the Right to the City*, London and New York: Routledge, 24–41.

Marshall, T.H. 1950. *Citizenship and Social Class*, Cambridge: Cambridge University Press.

Marx, Karl. 1963. *Early Writings*, T.B. Bottomore (ed. and trans.), New York: McGraw-Hill.

Marx, Karl. 2001 [1871]. *The Civil War in France*, London: Electric Book Company.

Massey, Douglas S. and Nancy A. Denton. 1993. *American Apartheid: Segregation and the Making of the Underclass*, Cambridge, MA: Harvard University Press.

Mastropaolo, Alfio. 2013. 'Le reinvenzioni del popolo', *Meridiana: Rivista di Storia e Scienze Sociali*, 77, 23–46.

Mayer, Margit. 2011. 'The "Right to the City" in Urban Social Movements', in Neil Brenner, Peter Marcuse and Margit Mayer (eds), *Cities for People, Not for Profit: Critical Urban Theory and the Right to the City*, London and New York: Routledge, 63–85.

McCormack, Derek P. 2008. 'Geographies for Moving Bodies: Thinking, Dancing, Spaces', *Geography Compass*, 2:6, 1822–1836.

McCormick, Carlo. 2011. 'The Writing on the Wall', in Jeffrey Deitsch (ed.) with Roger Gastman and Aaron Rose, *Art in the Streets*, New York: Skira Rizzoli Publications, 19–25.

Meegan, Rua and Lauren Teeling. 2010. *Irish Street Art: Stencils, Paste Ups, Murals & Portraits*, Dublin: Visual Feast Productions.

Menon, Harish C. 2016. 'India's Dalits Strike Back at Centuries of Oppression by Letting Dead Cows Rot on the Streets', *Quartz India*, 22 July. http://qz.com/738758/indias-dalits-strike-back-at-centuries-of-oppression-by-letting-dead-cows-rot-on-the-streets/

Mercille, Julien. 2014. 'The Role of the Media in Sustaining Ireland's Housing Bubble', *New Political Economy*, 19:2, 282–301.

Mercille, Julien. 2015. *The Political Economy and Media Coverage of the European Economic Crisis: The Case of Ireland*, London: Routledge.

Milkman, Ruth. 2012. 'Revolt of the College-Educated Millennials', *Contexts*, 11:2, Spring, 13–14.

Miller, Byron. 2006. 'Modes of Governance, Modes of Resistance: Contesting Neoliberalism in Calgary, in *Contesting Neoliberalism: Urban Frontiers*, New York: Guilford Press, 223–249.
Mitchell, W.J.T. 2013. 'Image, Space, Revolution: The Arts of Occupation', in W.J.T. Mitchell, Bernard Harcourt and Michael Taussig, *Occupy: Three Inquiries in Disobedience*, Chicago: TRIOS/University of Chicago Press, 93–129.
Moreiras, Alberto. 2004. 'Infrapolitics and Immaterial Reflection', *Polygraph*, 15/16, 33–46.
Morgan, Vanessa Sloan. 2014. 'Empty Words on Occupied Lands? Positionality, Settler Colonialism, and the Politics of Recognition'. *Antipode Foundation* https://antipodefoundation.org/2014/07/02/empty-words-on-occupied-lands/
Mouffe, Chantal. 1993. *Dimensions of Radical Democracy: Pluralism, Citizenship, Community*, London: Verso.
Mouffe, Chantal. 2005 [1993]. *The Return of the Political*, London and New York: Verso.
Mouffe, Chantal. 2009. *The Democratic Paradox*, London: Verso.
Mouffe, Chantal. 2013. *Agonistics: Thinking the World Politically*, London: Verso.
Mura, Virgilio. 2014. 'Paternalismo e democrazia liberale: un equivoco da chiarire', *Meridiana: Rivista di Storia e Scienze Sociali*, 79:1, 47–69.
Nagam, Julie. 2015. 'The Occupation of Space: Creatively Transforming Indigenous Living Histories in Urban Spaces', in Janine Marchessault, Chloë Brushwood Rose, Jennifer Foster and Aleksandra Kaminska (eds), *Land|Slide: Possible Futures*, Toronto: ABC Books.
Nash, Catherine J. and Alison L. Bain. 2007. '"Reclaiming Raunch"?: Spatializing Queer Identities at a Toronto Women's Bathhouse Event', *Social and Cultural Geography*, 8:1, 47–62.
Negri, Antonio. 2008. *The Porcelain Workshop: For a New Grammar of Politics*, Los Angeles: Semiotext(e).
Nepstad, Sharon Erickson. 2011. *Nonviolent Revolutions: Civil Resistance in the 20th Century*, New York: Oxford University Press.
O'Casey, Sean. 1924. *Juno and the Paycock*, London: Macmillan.
Oldenburg, R. (ed.) 2001. *Celebrating the Third Place: Inspiring Stories About The 'Great Good Places' At The Heart Of Our Communities*, New York: Marlowe and Co.
Ostroy, Andy. 2012. 'The Failure of Occupy Wall Street', *Huffington Post*, 31 May (updated 31 July). http://www.huffingtonpost.com/andy-ostroy/the-failure-of-occupy-wal_b_1558787.html
Parenti, Christian. 2015. 'The Environment Making State: Territory, Nature, and Value', *Antipode*, 47:4, 827–848.
Peake, Linda. 2016a. 'The Twenty-First-Century Quest for Feminism and the Global Urban', *International Journal of Urban and Regional Research*, 40:1, 219–227.
Peake, Linda. 2016b. 'On Feminism and Feminist Allies in Knowledge Production in Urban Geography', *Urban Geography*, 37:6, 830–838.
Peck, Jamie, Nik Theodore and Neil Brenner. 2012. 'Neoliberalism Resurgent? Market Rule after the Great Recession', *South Atlantic Quarterly*, 111:2, 265–288.
Porter, David (ed.) 1983. *Vision on Fire: Emma Goldman and the Spanish Revolution*, New Paltz, NY: Commonground Press.
Purcell, Mark. 2006. 'Urban Democracy and the Local Trap', *Urban Studies*, 43:11, 1921–1941.

Purcell, Mark. 2013. *The Down-Deep Delight of Democracy*, Oxford and Malden, MA: Wiley-Blackwell.

Ramachandran, V.R. 2011. *The Tell-Tale Brain: Unlocking the Mystery of Human Nature*, London: Heinemann.

Rancière, Jacques. 1999. *Disagreement: Politics and Philosophy*, Minneapolis, MN: University of Minnesota Press.

Rancière, Jacques. 2010. *Dissensus: On Politics and Aesthetics*, Steven Corcoran (trans. and ed.), London and New York: Bloomsbury.

Rancière, Jacques. 2014. *Hatred of Democracy*, Steven Corcoran (trans.), London: Verso.

Randolph, Brandi and Isabelle Bowker. 2016. 'Interview with Kenneth Morrison Wernsdorfer', *Poetry and Power*, Spring. http://www.poetryandpower.org/kenneth-morrison.html

Rankin, Katharine N. 2011. 'The Praxis of Planning and the Contributions of Critical Development Studies', in Neil Brenner, Peter Marcuse and Margit Mayer (eds), *Cities for People, Not for Profit: Critical Urban Theory and the Right to the City*, London and New York: Routledge, 102–116.

Rankine, Claudia. 2015. *Citizen: An American Lyric*. Minneapolis, MN: Greywolf Press.

Ratto, Matt and Megan Boler (eds) 2014. *DIY Citizenship: Critical Making and Social Media*, Cambridge, MA: MIT Press.

Reilly, Ian. 2014. 'Just Say Yes: DIY-ing the Yes Men', in Matt Ratto and Megan Boler (eds), *DIY Citizenship: Critical Making and Social Media*, Cambridge, MA: MIT Press, 127–136.

Ridgley, Jennifer. 2008. 'Cities of Refuge: Immigration Enforcement, Police, and the Insurgent Genealogies of Citizenship in U.S. Sanctuary Cities', *Urban Geography*, 29:1, 53–77.

Roberts, Alisdair. 2012. 'Why the Occupy Movement Failed', *Public Administration Review*, 72:5 (September/October), 754–762.

Rolston, Bill. 2004. 'The War of the Walls: Political Murals in Northern Ireland', *Museum International*, 56:3, 38–45.

Rosanvallon, Pierre. 2006. *La contre-démocratie: La politique à l'âge de la defiance*, Paris: Éditions du Seuil.

Rosanvallon, Pierre. 2014. *Le Parlement des invisibles*, Paris: Éditions du Seuil.

Rose, Gillian. 1993. *Feminism and Geography: The Limits of Geographical Knowledge*, Minneapolis, MN: University of Minnesota Press.

Rose, Nikolas. 1996. 'The Death of the Social? Refiguring the Territory of Government', *Economy and Society*, 25:3, 327–356.

Roy, Ananya. 2016. 'What is Urban about Urban Theory?' *Urban Geography*, 37:6, 810–823.

RTÉ News. 2010. 'Gardaí, Students Clash in Dublin', *RTÉ News*, 3 November. http://www.rte.ie/news/2010/1103/137628-education/

Ruggiero, Greg. 2012. 'Editor's Note', Noam Chomsky, *Occupy* (Occupied Media Pamphlet Series). New York: Zuccotti Park Press.

Runciman, Carin. 2016. 'Mobilising and Organising in Precarious Times: Analysing Contemporary Collective Action in South Africa', *International Journal of Sociology and Social Policy*, 36:9/10, 613–628.

Salek, Yasi. 2014. 'Audrey Wollen on Sad Girl Theory', *Cultist Zine*, 19 June, http://www.cultistzine.com/2014/06/19/cult-talk-audrey-wollen-on-sad-girl-theory/

Sassen, Saskia. 1997. 'Whose City Is It? Globalization and the Formation of New Claims', *Public Culture*, 82, 205–223.
Schmid, Christian. 2011. 'Henri Lefebvre, the Right to the City, and the New Metropolitan Mainstream', in Neil Brenner, Peter Marcuse and Margit Mayer (eds), *Cities for People, Not for Profit: Critical Urban Theory and the Right to the City*, London and New York: Routledge, 42–62.
Schneider, Nathan. 2013. *Thank You Anarchy: Notes from the Occupy Apocalypse*, Berkeley: University of California Press.
Scott, James C. 1985. *Weapons of the Weak: Everyday Forms of Peasant Resistance*, New Haven, CT: Yale University Press.
Scott, James C. 1990. *Domination and the Arts of Resistance: Hidden Transcripts*, New Haven, CT: Yale University Press.
Scott, James C. 1998. *Seeing Like a State: How Certain Schemes to Improve the Human Condition Have Failed*, New Haven, CT: Yale University Press.
Scott, James C. 2009. *The Art of Not Being Governed: An Anarchist History of Upland Southeast Asia*, New Haven, CT: Yale University Press.
Scott, James C. 2012. *Two Cheers for Anarchism: Six Easy Pieces on Autonomy, Dignity, and Meaningful Work and Play*, Princeton, NJ: Princeton University Press.
Scuccimarra, Luca. 2013. 'Il ritorno del popolo: Un'introduzione', *Meridiana: Rivista di Storia e Scienze Sociali*, 77, 9–21.
Sedgwick, Eve Kosofsky. 1997. 'Paranoid Reading and Reparative Reading; or, You're So Paranoid, You Probably Think This Introduction is about You', in Eve Kosofsky Sedgwick (ed.), *Novel Gazing: Queer Readings in Fiction*, Durham, NC: Duke University Press.
Shannon, D., A. Nocella and J. Asimakopoulos. 2012. 'Anarchist Economics: A Holistic View', in D. Shannon, A. Nocella and J. Asimakopoulos (eds), *The Accumulation of Freedom: Writings on Anarchist Economics*, Oakland, CA: AK Press.
Sharp, Deen and Claire Panetta. 2016. *Beyond the Square: Urbanism and the Arab Uprisings*, New York: Urban Research.
Sheehan, Helena. 2011. 'Occupy Dublin: Take Back the World they Have Stolen from us', *Irish Left Review*, 16 October. http://www.irishleftreview.org/2011/10/16/occupy-dublin-world-stolen/
Sieverts, Thomas. 2003, *Cities Without Cities: An Interpretation of the Zwischenstadt*, New York: Routledge.
Simpson, Audra. 2014. *Mohawk Interruptus: Political Life across the Borders of Settler States*, Durham, NC: Duke University Press.
Sisk, John & Son. 1987. *The Big Bank*, documentary film.
Sitrin, Miriam. 2014. 'The DNA of Occupy', *Popular Resistance* news website, 14 September. https://www.popularresistance.org/the-dna-of-occupy/
Smith, Laura Tuhiwai. 2002. *Decolonizing Methodologies: Research and Indigenous Peoples*, London and New York: Zed Books.
Smith, Michael Peter. 2001. *Transnational Urbanism: Locating Globalization*, London: Blackwell.
Smith, Michael Peter and Michael McQuarrie (eds) 2012. *Remaking Urban Citizenship: Organizations, Institutions and the Right to the City*, New Brunswick, NJ and London: Transaction Publishers.
Smith, Neil. 1987. 'Gentrification and the Rent Gap', *Annals of the American Association of Geographers*, 77:3, 462–465.

Smith, Neil. 1996. *New Urban Frontier: Gentrification and the Revanchist City*, New York: Routledge.

Snyder, Sharon L. and David T. Mitchell. 2001. 'Re-engaging the Body: Disability Studies and the Resistance to Embodiment', *Public Culture*, 13:3, 367–389.

Solnit, Rebecca 2013 'Foreword', in Nathan Schneider, *Thank You Anarchy: Notes from the Occupy Apocalypse*. Berkeley: University of California Press, ix–xii.

Spivak, Gayatri Chakravorty. 1988. 'Can the Subaltern Speak?' in Cary Nelson and Lawrence Grossberg (eds), *Marxism and the Interpretation of Culture*, Urbana: University of Illinois Press, 271–313.

Springer, Simon. 2010. 'Neoliberalism and Geography: Expansion, Variegation, Formations', *Geography Compass*, 4:8, 1025–1038.

Springer, Simon. 2014. 'Why a Radical Geography Must Be Anarchist', *Dialogues in Human Geography*, 4:3, 249–270.

Stansell, Christine. 1987, *City of Women: Sex and Class in New York, 1789–1860*, Urbana: University of Illinois Press.

Strange, Carolyn. 1995, *Toronto's Girl Problem: The Perils and Pleasures of the City, 1880–1930*, Toronto, ON: University of Toronto Press.

Tackett, Timothy. 2015. *The Coming of the Terror in the French Revolution*, Cambridge, MA: Belknap Press.

Takaki, Ronald. 1989. *Strangers from Another Shore: A History of Asian Americans*, New York: Little, Brown and Company.

Taussig, Michael. 2013. 'I'm So Angry I Made a Sign', in W.J.T. Mitchell, Bernard Harcourt and Michael Taussig, *Occupy: Three Inquiries in Disobedience*, Chicago: TRIOS/University of Chicago Press, 3–44.

Tufts, Steven and Mark Thomas. 2014. 'Populist Unionism Confronts Austerity in Canada', *Labor Studies Journal*, 39:1, 60–82.

United Nations Human Rights Advisory Panel. 2015. *Kadri Balaj, Shaban Xheladini, Zenel Zeneli and Mustafe Nerjovaj against UNMIK, Case No. 04/07, Opinion*, 27 February.

van Gelder, Sarah (ed.) 2011. *This Changes Everything: Occupy Wall Stree and the 99% Movement*, San Francisco: Berrett-Koehler.

Van Kempen, Ronald and Alan Murie. 2009. 'The New Divided City: Changing Patterns in European Cities', *Tijdschrift voor Economische en Sociale Geografie*, 100:4, 377–398.

Véron, Ophélie. 2016. '(Extra)Ordinary Activism: Veganism and the Shaping of Hemeratopias', *International Journal of Sociology and Social Policy*, 36:11/12, 756–773.

Vodovnik, Ziga 2013. *A Living Spirit of Revolt: The Infrapolitics of Anarchism*, Oakland: PM Press.

Walks, Alan. 2013. 'Mapping the Urban Debtscape: The Geography of Household Debt in Canadian Cities', *Urban Geography*. 34:2, 153–187.

Walks, Alan. 2014a. 'From Financialization to Urban Socio-Spatial polarization? Evidence from Canadian Cities', *Economic Geography*, 90:1, 33–66.

Walks, Alan. 2014b. 'Canada's Housing Bubble Story: Mortgage Securitization, The State, and the Global Financial Crisis', *International Journal of Urban and Regional Research*, 38:1, 256–284

We Are the Left. 2016. 'An Open Letter on Identity Politics, to and from the Left', *Medium*, 13 July. https://medium.com/@We_Are_The_Left/an-open-letter-on-identity-politics-to-and-from-the-left-b927fe66d3a4#.j57hz4569

Weaver, Jase, Craig S. Womack and Robert A. Warrior. 2006. *American Indian Literary Nationalism*, Albuquerque, NM: University of New Mexico Press.

White, Melanie. 2006. 'The Dispositions of Good Citizenship: Character, Symbolic Power and Disinterest', *Journal of Civil Society*, 22, 111–122.

White, Richard. J. 2009. 'Explaining why the Non-commodified Sphere of Mutual Aid is so Pervasive in the Advanced Economies: Some Case Study Evidence from an English City', *International Journal of Sociology and Social Policy*, 29:9/10, 457–472.

White, Richard J. and Colin Williams. 2014. 'Anarchist Economic Practices in a "Capitalist" Society: Some Implications for Organisation and the Future of Work', *Ephemera: Theory and Politics in Organization*, 14:4, 951–975.

White, Richard J. and Colin Williams. 2016. 'Beyond Capitalocentricism: Are Non-capitalist Work Practices "Alternatives"?' *AREA*, 48:3, 325–331.

Wilkerson, Isabel. 2010. *The Warmth of Other Suns: The Epic Story of America's Great Migration*, New York: Random House.

Williams, Dana. 2012. 'The Anarchist DNA of Occupy', *Contexts*, 11:2, Spring, 19–20.

Williams, Patricia. 1991. *The Alchemy of Race and Rights: Diary of a Law Professor*, Cambridge, MA: Harvard University Press.

Womack, Craig. 1999. *Red on Red: Native American Literary Separatism*, Minneapolis, MN: University of Minnesota Press.

Wood, Lesley. 2014. *Crisis and Control: The Militarization of Protest Policing*, London: Pluto Press.

Wood, Nicholas. 2006. 'Displaced Gypsies at Risk from Lead in Kosovo Camps', *New York Times*, 5 February. http://www.nytimes.com/2006/02/05/world/europe/displaced-gypsies-at-risk-from-lead-in-kosovo-camps.html?_r=0

Wood, Patricia. 2001. *Nationalism from the Margins*, Montreal and Kingston: McGill-Queen's University Press.

Wood, Patricia. 2013. 'Citizenship in the In-Between City', *Canadian Journal of Urban Research*, 21:1, 111–125.

Wood, Patricia. 2014. 'Urban Citizenship', in H. van der Heiden (ed.) *Handbook of Political Citizenship and Social Movements*, Cheltenham and Northampton, MA: Edward Elgar, 161–178.

Wood, Patricia and Rodney K. Brunson. 2011. 'Geographies of Resilient Social Networks: The Role of African American Barbershops', *Urban Geography*, 32:2, 228–243.

Wood, Patricia and Scot Wortley. 2010. 'AlieNation: Racism, Injustice and Other Obstacles to Full Citizenship', *CERIS: Ontario Metropolis Centre Working Paper series (August)*.

Writers for the 99% (ed.) 2012. *Occupying Wall Street: The Inside Story of an Action that Changed America*, New York: O/R Books.

Yiftachel, Oren. 2011. 'Critical Theory and "Gray Space": Mobilization of the Colonized', in Neil Brenner, Peter Marcuse and Margit Mayer (eds), *Cities for People, Not for Profit: Critical Urban Theory and the Right to the City*, London and New York: Routledge, 150–170.

Yingling, Thomas E. 1997. *AIDS and the National Body*, Robyn Wiegman (ed.), Durham, NC: Duke University Press.

Young, Douglas, Patricia Burke Wood and Roger Keil (eds) 2010. *In-Between Infrastructure: Urban Connectivity in an Age of Vulnerability*, Kelowna, BC: Praxis (e) Press.

Young, Iris Marion. 1990. *Justice and the Politics of Difference*, Princeton, NJ: Princeton University Press.

Young, Iris Marion. 1997a. 'Unruly Categories: A Critique of Nancy Fraser's Dual Systems Theory', *New Left Review*, 0.222 (March–April), 147–160.
Young, Iris Marion. 1997b. *Intersecting Voices: Dilemmas of Gender, Political Philosophy, and Policy*, Princeton, NJ: Princeton University Press.
Young, Iris Marion. 2000. *Inclusion and Democracy*, Oxford and New York: Oxford University Press.
Žižek, Slavoj. 1999. *The Ticklish Subject: The Absent Centre of Political Ontology*, London: Verso.
Žižek, Slavoj. 2011. 'Don't Fall in Love with Yourselves', in Carla Blumenkranz et al. (eds), *Occupy! Scenes from Occupied America*, New York: Verso, 66–69.

Index